The Internet

A Guide for Chemists

The Internet
A Guide for Chemists

Stephen M. Bachrach, Editor

Stephen R. Heller, Consulting Editor
Computer Applications in Chemistry Books

American Chemical Society
Washington, DC 1996

Library of Congress Cataloging-in-Publication Data

The Internet: A Guide for Chemists / Steven M. Bachrach, editor.
p. cm.
Includes bibliographical references and index.
ISBN 0–8412–3223–7 (clothbound). — ISBN 0–8412–3224–5 (pbk.)

1. Chemistry —Computer network resources. 2. Internet (Computer network).
I. Bachrach, Steven M., 1959–
QD8.3.I57 1995
004.6'7—dc20 95–41967
CIP

PRINTED IN THE UNITED STATES OF AMERICA

Advisory Board

Computer Applications in Chemistry Books

Contents

Specifically for Chemists

About the Editor

STEVEN M. BACHRACH is an associate professor of chemistry at Northern Illinois University. He received his B.S. in chemistry from the University of Illinois at Urbana in 1981 and a Ph.D. in organic chemistry from the University of California at Berkeley in 1985. He held a Director's Postdoctoral Fellowship at Los Alamos National Laboratory from 1985 to 1987 before joining the faculty at Northern Illinois University.

Bachrach's research interests include theoretical organic chemistry and the use of the Internet for disseminating chemical information. His efforts in theoretical chemistry have focused on interpretation of electron density and elucidation of reaction mechanisms, particularly pericyclic reactions or organophosphorus compounds and substitution reactions at heteroatoms. On the Internet, he has established a number of networked databases for chemists and has hosted many electronic conferences. Bachrach is a member of the American Chemical Society and Sigma Xi. He has published more than 50 papers. This is his first book.

Contributors

Steven M. Bachrach, Editor
smb@smb.chem.niu.edu
Department of Chemistry
Northern Illinois University
DeKalb, IL 60015

Charles F. Bender
bender@osc.edu
Ohio Supercomputer Center
1224 Kinnear Road
Columbus, OH 43212

Thomas J. Cozzolino
tcozz@rohmhaas.com
Rohm and Haas, Research Division
Routes 413 and 13
Bristol, PA 19007

Laurence R. Dusold
lrd@vm.cfsan.fda.gov
Center for Food Safety and Applied Nutrition
U.S. Food and Drug Administration
200 C Street, S.W.
Washington, DC 20204

Thomas C. O'Haver
Thomas_C_O'HAVER@umail.umd.edu
Department of Chemistry and Biochemistry
University of Maryland
College Park, MD 20742

Jan K. Labanowski
jkl@osc.edu
Ohio Supercomputer Center
1224 Kinnear Road
Columbus, OH 43212

Thomas H. Pierce
THPierce@rohmhaas.com
Rohm and Haas, Research Division
Routes 413 and 13
Bristol, PA 19007

Alejandro Pisanty
apisan@servidor.unam.mx
Facultad de Quimica y Comite Asesor de Computo
Universidad Nacional Autonoma de Mexico
Ciudad Universitaria, 04510 DF, Mexico

Henry S. Rzepa
h.rzepa@ic.ac.uk
Department of Chemistry
Imperial College of Science, Technology, and Medicine
London, SW7 2AY, United Kingdom

Frank A. Tinker
tinker@physics.arizona.edu
Zetetic Institute
1665 East 18th Street
Tucson, AZ 85719

Preface

PERUSE THE GENERAL MEDIA, newspapers, weekly magazines, even television, and you cannot help but come across an overload of hype about the coming "information revolution", the "electronic superhighway", and the "infobahn". The U.S. Government considers a national initiative to connect all of its citizens. The media giants discuss plans to offer 500 cable television channels to the home, with the user selecting the programming. Interactive televideo is proposed in television commercials, suggesting a future in which employees will work at home and connect through their phone or cable lines to their offices.

Given this hype and interest in the future, what is the current state of computer networks and information distribution? The existing global computer network is the Internet. Developed initially by the U.S. military to promote communications between its myriad facilities, the Internet has grown into a heterogeneous mixture of computer platforms and software, connecting academic, government, and commercial sites around the world, reaching even into war-torn countries and Third World nations.

Concerning the Internet, once again, the mixture of hype and information is staggering. On a recent visit to a local bookstore, I found nearly seven bookshelves filled with guides, manuals, and how-to books on various aspects of the Internet! One could find a yellow pages of the Internet, introduction guides to the World Wide Web (WWW), how to use `Mosaic`, and even technical manuals on TCP/IP.

Given this background, what is the purpose of this book, *The Internet: A Guide for Chemists*? Yet another entry to the Internet publication derby? The vast majority of popular books on the Internet are geared toward the general user and designed to guide this person through the tasks necessary to access information on the Internet. These books will show the user how to connect to the Internet; how to use the major Internet resources, such as electronic mail, Gopher, and the WWW; and where to find general types of information, such as the location of the White House Home Page and databases on AIDS.

The purpose of this book is to focus on the Internet resources now available for chemists. We present a description of a large number of different resources, such as material safety data sheets (MSDS), commercial products, graduate school programs, and electronic conferences. We describe how to use the major tools to access these resources, with enough detail to enable the beginner to get to the sites, along with tips that will likely be new and useful even to the power users.

But perhaps even more important is our desire to encourage more individuals to become information providers on the Internet. This feature is what most sets the aims of this book apart from the popular media guides. We present detailed information on how to set up your own Gopher and WWW sites, including where to obtain the necessary software, how to build the server, and how to maintain the site. We describe some of the decisions and guidelines for establishing your own database and presenting it on the Internet. A chapter is devoted to how to plan, organize, and manage an electronic conference, allowing scientists from around the world to participate in a scientific meeting from their own offices. The last chapter details how one company has designed and integrated Internet resources within the corporate environment, although the ideas presented can be generalized to any organization.

We have tried to write our chapters with the assumption that the reader is a novice to the Internet, and we have had to make a few other assumptions. We do not address how to get connected to the Internet. So many different computer platforms, operating systems, and applications are being used, and there are many different Internet providers, so that we cannot begin to address these issues within the scope of this book. If you are not connected to the Internet, you will need to contact your system administrator or Internet provider to obtain instructions on how to configure your system and how to connect.

The great hope for the Internet is that it will continue to grow because more and more useful and valuable information will continue to be offered by this medium. We hope that this book will spur others to develop their own Internet resources and create exciting, novel applications that will entice the chemical community.

I thank my wife, Carmen, for her help and support during the preparation of this book, and always.

Steven M. Bachrach
Department of Chemistry
Northern Illinois University
DeKalb, IL 60015

May 1995

part one

Basic Information

chapter one

History of the Internet

Laurence R. Dusold

In This Chapter

This chapter gives a description and brief history of the Internet and instructions for connecting to it.

Published 1996 American Chemical Society

What Is the Internet?

The Internet is a matrix of networks and computer systems linked together around the world.[1] Networks have no political borders or boundaries on the exchange of information. Networks are connected by gateways that effectively remove barriers so that one type of network can "talk" to a different type of network. The Internet is the largest of these different networks in terms of the number of sites interconnected and the number of users.

In this chapter, the Internet's history and evolution into what today is popularly known as the information superhighway will be described. Also, the protocol running on the Internet, the transmission control protocol/Internet protocol (TCP/IP),[2–5] will be described informally in the context of its evolution in the history of the Internet.

The most important reason for all of the interest in the Internet is not the network itself or how it works technically, but the information and data exchange available to the users. The explosive growth of the Internet can be seen in Figure 1.1.[6]

ARPANET

The First Network

In 1969 the Defense Advanced Research Projects Agency (DARPA) funded the establishment of a few connected sites. These first few sites eventually evolved into the ARPANET. The military intent was to build a network that continued to work in the event of a war if parts of the network were broken. DARPA's goal in this original network design was to show the feasibility of a packet-switched network that would have no single point of failure. The "packet" contained a short piece of electronic data or information that would later be put back together with other packets at the receiving computer into the complete message. All of the connection points in the network were equal, with no central administration. Paul Baran described what was to become the ARPANET in a Rand Corporation paper[7] that was made public in 1964.

By the end of 1969 the first four sites were running on the infant ARPANET. These sites were at the University of California at Los Angeles, Stanford Research Institute, the University of California at

Figure 1.1. NSFNET byte traffic from March 1991 to September 1994.

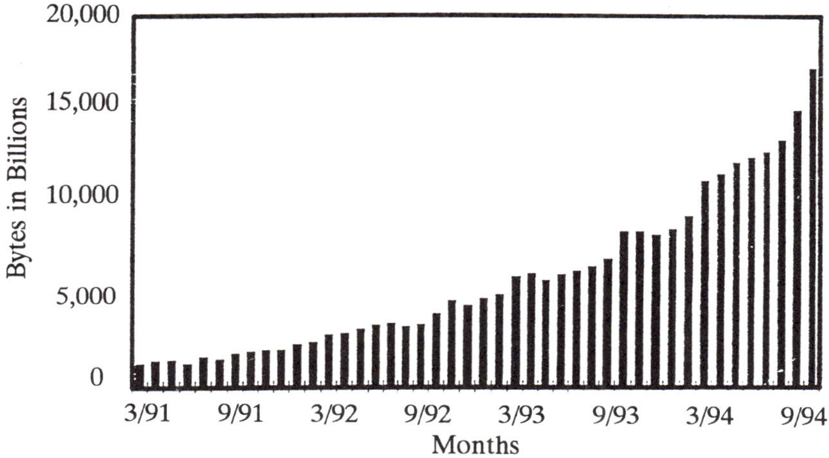

Santa Barbara, and the University of Utah. These four research sites had computer centers that were doing network research. Each site had an interface message processor (IMP) computer connected to the network. The IMP computers were Honeywell minicomputers with 12K (K is kilobyte) of memory, which is very small when compared with today's standard personal computer (PC) on a desktop. The IMP computers broke down the message traffic from the sending site into packets, and at the receiving site another IMP assembled the packets back together. The original speed of the links between the four sites was 50,000 bits per second (or 50 Kbps; here K means thousand). The original network research studied different ways the network could fail when it got too many fragments of messages at one time. (This was later called network congestion).

The success of the beginning ARPANET soon became apparent when it started to grow beyond the original four sites. In 1971 15 interconnected sites were using the IMP minicomputers. By 1972 the ARPANET had increased to 37 sites. An interesting discovery was made during these early beginnings of the ARPANET. Although the U.S. Department of Defense (DOD) had intended the network to be used mainly for computer research into making robust networks and for remote computing between sites, the users also started to exchange messages with one another because it was so convenient. The research exchange was getting done, but the volume of traffic devoted to personal messages and what in the future would be called electronic mail (e-mail) greatly exceeded anyone's imagination.

Request for Comments (RFC)

During the very early stages of the development of the ARPANET there were no network standards, or universally agreed upon ways of doing things between sites. So, an informal process evolved, the request for comments or RFC. Fortuitously, the RFC process helped to accelerate the growth of the Internet. The draft RFCs would suggest possible standards or ways of communicating and exchanging information, using software and protocols that everyone could probably agree upon. They were published as drafts, commented upon, and finally accepted as networking standards. When the draft RFC became final it would be assigned a number so that it could be referenced as an Internet Standard.

The first RFC was published in 1969 by Steve Crocker.[8] Today there are about 2000 RFCs.[9] Because draft standards were proposed first and feedback was requested from everyone on the network, the standards became more widely accepted by everyone participating in the evolving Internet. Today the RFCs have become the de facto network standards that every major vendor must follow to sell network hardware and software.

A series of RFCs grouped together, known as the for-your-information (FYI), covers topics on everything from getting started to ethics. The titles of the current FYIs available on the Internet are summarized in Appendix 1.

Protocols for Networking

Some of the computers that were being interconnected in the early 1970s on the ARPANET were Honeywell, IBM, DEC, and Xerox Data Systems. Any computer that happened to be at one of the participating sites might get "connected". No specific hardware or brand was required, other than the IMP computer to connect to the other computers in the expanding ARPANET. In October 1972 the first public demonstration of the young ARPANET took place at the International Conference on Computer Communications held in Washington, DC.

The public demonstration worked better than the organizers had imagined.

During this time frame, the first electronic mail, or e-mail, program was written, and the concept of e-mail distribution lists (today's list servers) also evolved. At this stage TCP/IP had not been invented, so it was not being used yet for the growing ARPANET. A network control program (NCP) was being used, but it had many limitations by today's TCP/IP internetwork standards. In May 1974 Vinton Cerf and Robert Cahn wrote the first paper on the emerging TCP protocols, entitled, "A Protocol for Packet Network Internetworking", published by *IEEE Transactions on Communications*. RFCs soon followed on the emerging TCP protocol. TCP is described in RFC-793.[10] TCP allows for the correction and error-free sending and receiving of packets of data on a computer network.

The Internet protocol (IP) is concerned with the transmission of packets between the source and destination. IP is not concerned with error recovery, just the successful movement of the packets from the originating site to the destination through the network. IP is described in RFC-791.[11]

TCP and IP are independent of the physical medium that they can run on. TCP/IP can be run on many other media, but the most popular are twisted-pair copper wire and fiber backbone cabling in buildings and campuses.

The operation of the ARPANET was moved to the Defense Communications Agency in 1975. In August 1975 the American Chemical Society Division of Computers in Chemistry sponsored a symposium on Computer Networking and Chemistry at the 170th national meeting in Chicago, Illinois. The ACS published the proceedings of this symposium in the ACS Symposium Series as No. 19.[12] The ARPANET was being used by chemists for chemical research.

The UNIX operating software was developed at the University of California at Berkeley, and it included networking software. One of the aspects that made UNIX very popular was the transportability of the software. In 1977 UNIX was distributed by AT&T Bell Labs, and in 1979 Usenet News was started using the new UNIX software, between Duke University and the University of North Carolina. Usenet News can be loosely described as a worldwide bulletin board system where anyone on the Internet can read any posting, and anyone else can reply to any posting with an answer or comment. Because

Usenet News is so "open" and unrestricted, it is one of the top three in network traffic on the Internet today.

Bitnet

In 1981 Bitnet started at the City University of New York with a connection to Yale University. Although the Bitnet network started as a network job entry (NJE) protocol network only, its e-mail contribution to the ARPANET through gateways was significant in the 1980s. These and other described networks are important because their data would be carried by the future Internet, and that would in turn fuel the further expansion of the ever-growing Internet.

CSNET (Computer Science Network) was also formed in 1981. In 1983 the European Academic and Research Network (EARN) was established, and it operated the same as the Bitnet network in the United States, to which it was also interconnected. EARN is linked to Bitnet as a cooperative network.

In 1989 CSNET merged with Bitnet. By 1994 Bitnet evolved into Bitnet–II and used TCP/IP as the connection protocol over the Internet. Bitnet–II designated a little over a dozen "core nodes" that electronically connected other Bitnet members in the United States. All of the Bitnet traffic is encapsulated in TCP/IP using special software, VMNET, written at Princeton University.

Today most of the Bitnet traffic is carried on the Internet. Bitnet, which is run by the Corporation for Research and Education Networking (CREN), has more than 2000 host computers in more than 30 countries. Bitnet host computers are usually larger computers (i.e., IBM or DEC mainframes, not PCs or workstations) having up to several thousand users per host at the larger sites. The Bitnet–II traffic on the Internet has consistently ranked in the top 14 by byte volume. It peaked at more than 170 billion bytes in one month in 1994 when this chapter was being written. Figure 1.2 shows a world map diagram of the Bitnet network.

Internet Starts Running

In 1982 the Defense Communications Agency (DCA) and DARPA established that TCP/ IP was to be the standard connection protocol

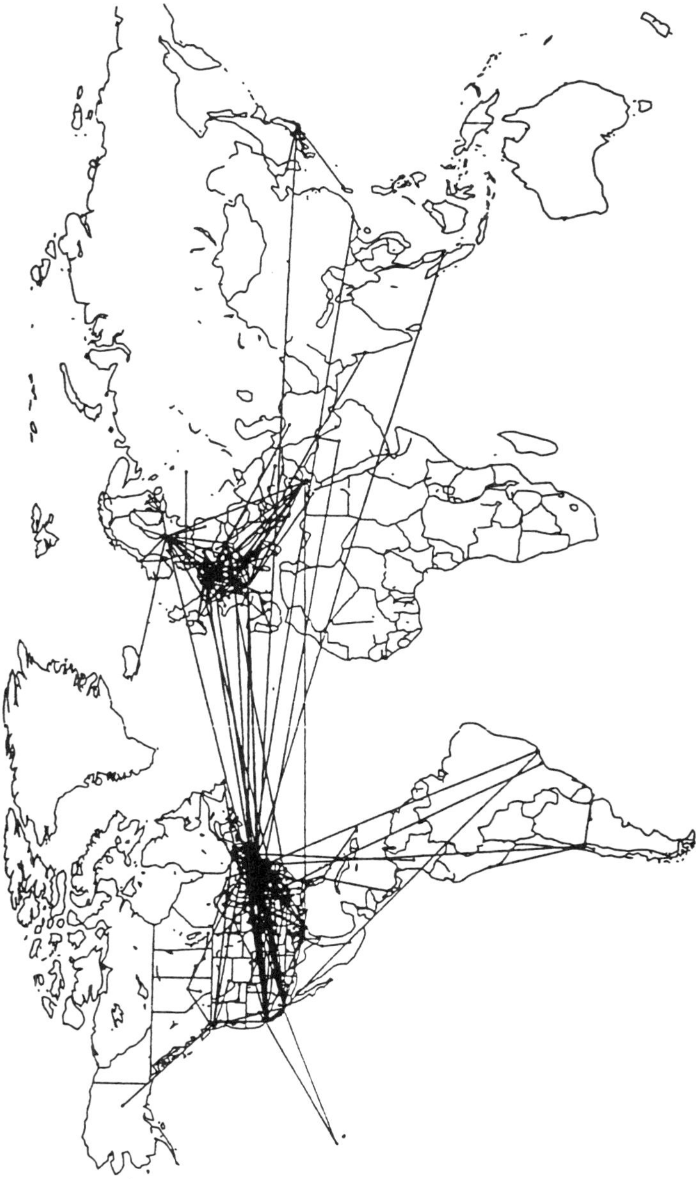

Figure 1.2. Bitnet connections worldwide. All published links are shown. Map data are taken from World Date Bank II Gall stereographic projection.

used on the ARPANET. DOD established the Defense Data Network as the umbrella network for all of the DOD during the year. The conversion to TCP/IP on the ARPANET was completed by January 1, 1983, to the new protocols, and the term "Internet" came into use from the term "the Internet protocol" (IP). In 1983 gateways connecting other networks, such as Bitnet, running protocols different from TCP/IP started to appear. Thus other networks were being connected to the ARPANET, and the ARPANET was evolving into a major carrier of TCP/IP traffic across the United States.

This tremendous growth caused a major event to happen in 1983: the ARPANET split into two networks, MILNET and ARPANET. For many this is the point that marks the beginning of the true Internet, because non-DOD sites were allowed to connect to the ARPANET after the MILNET split from the ARPANET, and TCP/IP was the accepted protocol running on the network.

The next major milestone in the evolution of the ARPANET occurred in 1984 when the number of interconnected hosts finally exceeded the 1000 mark. That same year the first domain name server (DNS) was put on the ARPANET. This server relieved users of the need to know the exact IP numbers and routes from one system to another. Also in 1984 national networks were started in Japan (JUNET or Japan UNIX Network) and in the United Kingdom (JANET or Joint Academic and Research Network).

NSFNET

The next significant event, was in 1986 when the NSFNET finally came into being. A network connecting five supercomputer centers was established by the National Science Foundation (NSF) to enhance computer and network research capabilities in the United States. NSFNET initially started with link speeds of 56 Kbps between the sites. The connection computers initially were LSI-11 computers from DEC that performed work on the network packets and other network functions that were evolving. The five sites were at the John von Neuman Computer Center at Princeton University, Pittsburgh Supercomputer Center at the University of Pittsburgh, the San Diego Supercomputer Center at the University of California at San Diego, the National Center for Supercomputer Applications at the University of Illinois at Urbana–Champaign, and the Cornell Theory Center at

Cornell University. NSFNET allowed widespread connection of regional networks and universities.

In 1987 NSF signed a cooperative agreement with Merit Network, Inc., allowing Merit to manage the NSFNET and to upgrade the network, because it was already becoming overloaded. By the summer of 1988 Merit, in cooperation with IBM providing high-speed routers and MCI providing the high-speed circuits, was able to upgrade the initial 56-Kbps NSFNET to a T-1 (1.544 million bits per second) speed network The number of computers connected to the ARPANET and NSFNET now exceeded 10,000 for the first time, and a period of very rapid expansion of these and other interconnected networks was occurring. The T-1 Internet has interconnections in the United States to the National Aeronautics and Space Administration (NASA) Science Internet (NSINET) and the Department of Energy's Energy Science Network (ESNET), in addition to about 500 different computer networks.

The Internet Worm

In May 1988 Clifford Stoll wrote a paper, "Stalking the Wily Hacker",[13] in which he showed that what had started as his attempt to track down a billing error in 1986 turned into an almost 2-year project. He traced an individual who had gotten into more than 30 computers on the Internet. On November 2, 1988, about 6000 computers on the Internet came under attack by a software program known as a "worm". For the first time, major parts of the network "broke" due to a security problem. Everyone was shocked at how quickly the worm affected only those sites running some variation of the UNIX operating system. Both events helped make it clear to the growing networks that network security should be considered and made a part of future expansion. Even with the negative publicity in 1988, it took only 2 years from the time the networks passed the 10,000 mark of connected computers in 1987 to exceed the 100,000 connected computers in 1989.

In June 1989, a special edition of *Communications of the ACM*, devoted in its entirety to the Internet Worm incident, carried a paper by Eugene Spafford entitled "The Internet Worm Crisis and Aftermath".[14] This special edition chronicled the events around the worm, how it was handled, and what should be taught to students about ethics related to the Internet. The topics covered in this special edition are relevant today on the Internet as well. Estimates at the time suggested that no more than 6000 computers running UNIX were affected by the worm, or very roughly less than 10% of the Internet.

One of the interesting points made was that the Internet survived this attack by the worm so well because there was diversity in the Internet. Not all of the network computers were supplied by one vendor, they were not all running the same operating system, and they used different implementations of TCP/IP. The vast number of different brands and different implementations actually was a strong point in the assessment of the aftermath of the worm. The hardware and software diversity in fact contributed to the robustness of the Internet to survive attacks based upon a given hardware or software bug or loophole.

Nevertheless, because of these security incidents, RFC-1087, "Ethics and the Internet", was published.[15] Also, as a direct result of these and other incidents, most judicious sites today run antivirus software on PCs and operate their TCP/IP networks behind "firewalls".[16] Firewalls are computers and software that provide a way for an organization to put a barrier between itself and the Internet in order to keep out unwanted accesses and to enforce some type of access policy. Firewalls can significantly increase the complexity of the network for end users but improve the security of operating a corporate or a university network attached to the Internet. Scientists operating behind firewalls should not become lax in their concern for network security, because firewalls do not provide absolute security, but just increase the level of network security of the university or corporation using them.

NREN, the National Research and Education Network

Other events of interest in the evolution of the Internet that happened in 1989 were the formation of the Internet Engineering Task Force (IETF) under the Internet Activities Board (IAB), and the publication of "The Federal High-Performance Computing Program" by the Executive Office of the President, Office of Science and Technology, on September 8, 1989. In this document the National Research and Education Network (NREN) was formally proposed and a time line established for the next decade of the NSFNET and the Internet in the United States.

The Congress of the United States, Office of Technology Assessment, also published a background paper in September 1989 entitled, “High-Performance Computing and Networking for Science”. Thus, both the Executive and Legislative branches of the U.S. Government were in basic agreement with the concept of establishing an NREN in the United States.

In 1991 the Gore Bill was passed by Congress and signed by the President on December 9. The purpose of this legislation, the High-Performance Computing Act of 1991, was to connect all schools, universities, and government agencies on one network. It established the development of a National Research and Education Network as part of its provisions.

NSFNET Expansion

In 1990 the National Science Foundation published its program guideline NSF–90–7, which promoted the continued expansion of the Internet. This program allowed any college or university in the United States to connect with the NSFNET. An excerpt from the NSF guideline follows:

> The NSFNET Program supports high-speed, wide-area computer communication networks with a goal of establishing an Internet or network of networks with three levels: a national backbone network, mid-level networks usually based around some geographical region of the country, and campus or local networks at educational and research institutions. The individual institutions are connected to a mid-level network in the appropriate geographical region. The mid-level network is in turn attached to the high-speed national backbone network, usually at its network operation center. The backbone is connected to other national networks including the Defense Research Internet, NASA Science Network, and the Energy Sciences Network; these interconnected networks and many others worldwide comprise the Internet....
>
> The gateway to the NSFNET should be available to all researchers, faculty, and scholars at the institution. Ideally the

institution will have installed a high-speed campus network and have adopted the TCP/IP protocols as standard.

This announcement was significant because no restrictions were placed on who could be connected as long as the Acceptable-Use Policy (AUP) of the National Science Foundation was followed. Generally under the AUP, any use that enhances education or the goals of the educational research community is allowed. Also, research and education users can use NSFNET to correspond with commercial companies and receive commercial services and products electronically. However, for-profit companies cannot use the backbone for advertising or other commercial purposes for which the companies initiate the contact. Universities and research organizations did not need to have any tie into computer network research or other government-funded projects of any type in order to connect to the NSFNET backbone or to the regional network to the NSFNET backbone in the United States.

Searching the Internet

In 1990 the ARPANET was officially decommissioned and ceased to exist. The other major occurrence in 1990 was that the software to search throughout the Internet for the location of electronic files emerged. This software, named Archie, is used to find files on the Internet so that users can use file-transfer protocol (FTP) to get a copy of the desired software. Archie was written at McGill University in Canada by Peter Deutsch, Alan Emtage, and Bill Heelan. Archie is now also offered as a commercial product with appropriate support available.

The regional networks throughout the United States allowed smaller colleges and universities to connect to the rapidly growing Internet. Figure 1.3 is a U.S. map of regional interconnection points in 1991. These providers serve local universities and industrial sites with research facilities in the United States.

The Commercial Internet Exchange (CIX) Association, which is a nonprofit trade association of public data internetwork providers, was formed in 1991 with for-profit members. The CIX had interconnects with the regional networks, and each carried the other's traffic without adding surcharges or traffic-based charges between CIX members. By joining the CIX, all members were assured that no restriction would be placed on the type of traffic routed between the member networks.

Figure 1.3. NSFNET connection points in August 1991. Different shaped symbols represent regional and local network providers who acted as intermediates.

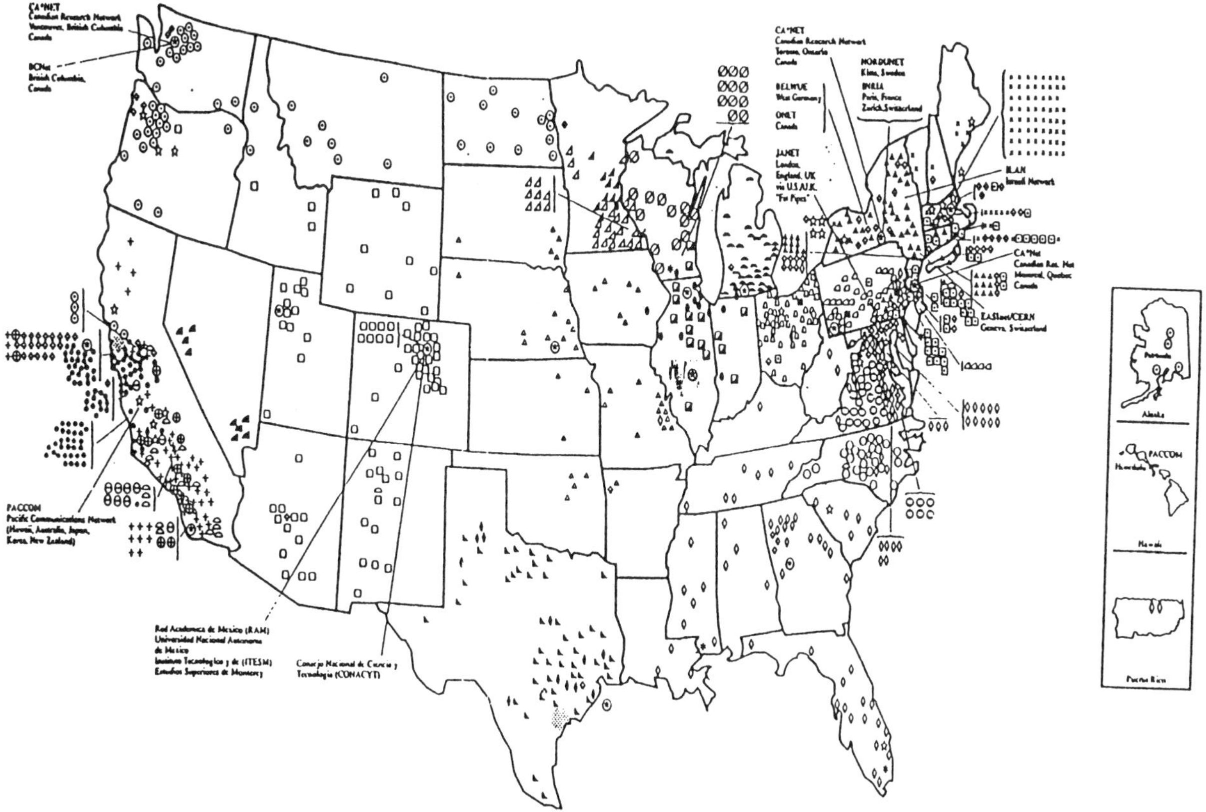

Thus there was no fear of violating the NSFNET acceptable-use guidelines, which banned commercial Internet traffic from the government-funded NSFNET.

In addition to the for-profit organizations getting together in 1991, the other significant events were that Gopher software, written by Paul Lindner and Mark McCahill, was released by the University of Minnesota, and WAIS (Wide Area Information Service) software, written by Brewster Kahle, was released on the Internet.

Gopher software allowed Internet users to use a client program running on their local computers or workstations to go to a Gopher server and review ordered lists of information. Gopher made such an impact that client software soon existed for everything from IBM mainframes, DEC VAXes, PCs running both DOS and Windows, and UNIX workstations. Gopher within 2 years would be contributing to the Internet traffic at the compound growth rate of almost 1000%! Gopher traffic on the NSFNET backbone in the United States was 374 billion bytes in January 1994, and it increased to 864 billion bytes by October 1994. Gopher will be discussed in detail in a later chapter in this book, because it is a tool that contributes significantly to the ease of getting useful information from the Internet.

WAIS provides the network user a way to search through collections of indexed data on the Internet. WAIS is a text-searching system that is distributed over the Internet, and it works because other users have indexed the words occurring in a document. When a user requests a WAIS search, the client contacts all known servers that have relevant information and gives back results ranked in order of most likely to match to least likely to match the search request.[17]

Because of this continued growth, the NSFNET was upgraded again in 1992. This time the upgrade was to a national T-3 (45 million bits per second) backbone network. The applications that were evolving on the Internet, and thus the NSFNET backbone in the United States, were helped by the upgrades and in turn fueled further larger increases in the amount of network traffic being carried by the NSFNET.

Other events in the history of the Internet that occurred during 1992 were that the number of host computers exceeded the 1,000,000 mark, the Internet Society (ISOC) was formed, and the Internet Activities Board (IAB) became a part of the ISOC. The Internet Society was formed as an international professional organization to foster the evolution and use of Internet technology, standardization, etc. The Society currently also publishes bimonthly *On the Internet* (ISSN 1081–3969) and sponsors international meetings on the Internet.[18]

The other event that began in 1989 and finally became public in 1992 was the first public distribution of the World Wide Web (WWW) client software on the Internet from CERN, the European Laboratory for Particle Physics, which is the home of the Web, in Geneva, Switzerland. The World Wide Web concept was designed and implemented by Tim Berners-Lee at CERN.

A WWW ***client*** is computer software that runs on a PC, workstation, or on the userid of a mainframe computer. This software is concerned with receiving and presenting the data and information to the user. Examples of WWW client programs are `Netscape`, `Mosaic`, `Lynx`, and `Charlotte`. `Netscape` runs on PCs and `Charlotte` runs on mainframes. A WWW ***server*** also is computer software that runs on PCs, workstations, or mainframes, but it is concerned with information retrieval on behalf of the WWW clients that request the data. This software has a collection of files, data, possibly graphics, and even multimedia. Upon receiving a request from a WWW client program, the server gives out the data from its computer system onto the Internet in a very standardized format. The standard format for data and information dissemination on the Internet is HyperText transport protocol (HTTP).

In 1993 the Internet Network Information Center (InterNIC) was started with the funding of the NSF to provide information, registration, and directory services for the Internet, itself, including all the standards and RFCs. NSF's awardees were General Atomics (CERFNET), Network Solutions, Inc., and AT&T.

Also in February 1993, the National Center for Supercomputing Applications (NCSA) in Illinois (Marc Andreeson's code) released the first graphical viewer, `Mosaic`, for the WWW.

Because of the graphical user interface (GUI) and the ability to display images, play sounds, and get text information, the growth of the World Wide Web has been phenomenal. In April 1994 the NSFNET backbone traffic from the WWW surpassed the Gopher. It is increasing at a rate that is unbelievable, until one looks at the NSFNET traffic from WWW. In January 1994 it was 269 billion bytes, and by October 1994 it had increased to 2.152 trillion bytes! In terms of total bytes on the NSFNET only file-transfer protocol (FTP) and network news transport protocol (NNTP), which supports Usenet News, are currently higher.

Future of the Internet

Because of this continued, nonstop growth of the Internet, in February 1994 the NSF formally announced the next stage for NSFNET, which had a major impact on the Internet in the United States at the end of May 1995, when the T-3 NSFNET backbone was decommissioned. An excerpt of a press release from February 14, 1994, is given in Appendix 2.[19] NSF's vision for the future was described at a briefing by NSF to explain the material in the press release.

Clearly, the Internet community is in the midst of a revolution in computer technology made possible by using workstations, PCs, mainframes, and supercomputers all tied together, with the Internet as the "glue". The very high speed backbone in the United States will also make feasible new applications that today are beyond imagination.

Uniform Resource Locators (URLs)

The URL or uniform resource locator originated with the World Wide Web project and Tim Berners-Lee. It allows any Internet resource to be specified and tells the reader how to get the information and where the resource is located. For example,

- URL **http://vm.cfsan.fda.gov** is a WWW server at a government Internet site, the Food and Drug Administration (FDA, Center for Food Safety and Applied Nutrition).
- URL **tn3270://ibm.rhrz.host.de** is an IBM mainframe in Germany and TN3270 (Telnet) is used to access it.
- URL **ftp://vax8.cfsan.fda.gov** is an anonymous FTP site at the FDA.
- URL **gopher://is.internic.net:70** is a Gopher site at the InterNIC.

If you are using a WWW client, a useful starting point to search for information on the Internet is at URL **http://vm.cfsan.fda.gov/referenc.html.** More details are given in Chapters 4 and 11.

How To Get Connected

If you want to start out with e-mail and a limited amount of Gopher, Usenet News, and World Wide Web access time per month on the

Internet, several choices are available with Windows-based or MAC-based (Apple Macintosh) graphical user interfaces. The obvious big three are CompuServe (1–800–487–9197), America Online (1–800–827–6364), and Prodigy (1–800–776–3449). The big three providers all give 5 or more hours per month access to the World Wide Web and essentially unlimited e-mail messages per month for less than $10. If you are affiliated with a small college or university that is not yet connected, a good starting point is the book *Internet: Getting Started.*[20] All of the names, addresses, and phone numbers of the regional networks and major international network providers are listed.

If you are interested in personal access, with more access hours than the big three provide, many software packages come with a "Network Provider" in the box. Before signing up with one of these services, you should fully understand all of the connection costs, telephone long-distance charges, and so on. For many Internet users, being restricted to a base number of "prime time" hours and unlimited evening or weekend use is the only choice because of the costs. Two dial-up providers worth checking out (at the time this was written) are NetCruiser from Netcom (1–800–353–6600) and the Pipeline, now sold nationally by Performance Systems International, Inc. (1–800–827–7482).

An excellent starting place for PC-based Internet dial-up access from home or office is *Connecting to the Internet: An O'Reilly Buyer's Guide.*[21]

Furthermore, more than a dozen FreeNets are now operating in North America; they provide e-mail access and some other, sometimes limited Internet, services.[22] You are asked to agree not to use the FreeNet for any commercial purposes and to abide by the local fair-use rules that may be in effect in order to make the resources available to as many users as possible. An additional source of general information on the Internet and its resources can be found in reference 17.

References

[1]Quarterman, John S. *The Matrix Computer Networks and Conferencing Systems Worldwide;* Digital Press: Bedford, MA, 1990.

[2]Stevens, W. Richard *TCP/IP Illustrated, Volume I;* Addison-Wesley: Reading, MA, 1994.

[3]Hunt, Craig *TCP/IP Network Administration;* O'Reilly & Associates: Sebastopol, CA, 1992 (January 1994 printing).
[4]IBM Networking Systems Redbooks CD-ROM SK2T–6022–02, which includes TCP/IP Tutorial and Technical Overview and TCP/IP for VM, OS/2, and MVS, plus other TCP/IP material. August 1994. IBM Corporation, Dept. E15, Research Triangle Park, NC 27709. (May be ordered from 1–800–879–2755.)
[5]Washburn, K.; Evans, J. T., *TCP/IP Running a Successful Network;* Addison-Wesley: Reading, MA, 1993.
[6]The numbers from various sources are at URL **http://vm.cfsan.fda.gov:80/~lrd/**. This URL presents a list of files available for view, including most of the full-text articles used in this chapter as references.
[7]URL **gopher://gopher.isoc.org:70/11/Internet/history/**. See also URL **ftp://umcc.umich.edu/pub/seraphim/doc/nethist8.txt**
[8]URL **ftp://ds.internic.net/rfc/**
[9]URL **ftp://ds.internic.net/rfc/**
[10]Transmission Control Protocol, DARPA Internet Program Protocol Specification, RFC-793, available on WWW at URL **ftp://ds.internic.net/rfc/ rfc793.txt**
[11]Internet Protocol, DARPA Internet Program Protocol Specification, RFC-791, available on WWW at URL **ftp://ds.internic.net/rfc/rfc791.txt**
[12]*Computer Networking and Chemistry;* ACS Symposium Series 19; Lykos, Peter, Ed.; American Chemical Society: Washington, DC, 1975.
[13]Stoll, Clifford "Stalking the Wily Hacker", *Commun. ACM* **1988,** *31,* (5), 484.
[14]Spafford, Eugene "The Internet Worm Crisis and Aftermath", *Commun. ACM* **1989,** *32 (6),* 678–687.
[15]"Ethics and the Internet", RFC-1087, available on WWW at URL **ftp://ds.internic.net/rfc/rfc1087.txt**
[16]Cheswick, William R.; Bellovin, Steven M. *Firewalls and Internet Security: Repelling the Wiley Hacker;* Addison-Wesley: Reading, MA, 1994 (second printing). Also, TIS, the Web server of Trusted Information Systems, Inc., gives many good examples and explanations of Internet security. Thus it is a good starting point to begin investigating security after reading the Cheswick et al book. It is accessible on the WWW at URL **http://www.tis.com**
[17]Krol, Ed *The Whole Internet: User's Guide and Catalog;* O'Reilly & Associates: Sebastopol, CA, 1994, second ed.
[18]URL **http://www.isoc.org** or URL **gopher://gopher.isoc.org**
[19]URL **http://www.nsf.gov**, which allows searching for all NSF reports and documents on NSFNET.
[20]Marine, April; Kirkpatrick, Susan; Neou, Vivian; Ward, Carol *Internet: Getting Started;* PTR Prentice Hall: New York, 1993.

[21]Estrada, Susan *Connecting to the Internet, An O'Reilly Buyer's Guide;* O'Reilly & Associates: Sebastopol, CA, 1993.
[22]URL **http://herald.usask.ca/~scottp/free.html**

Appendix 1: FYI Index

This file contains citations for all FYIs in numeric order. Details on obtaining FYIs via FTP or e-mail may be obtained by sending an e-mail message to: **rfc-info@ISI.EDU** with the message body `help: ways_to_get_fyis`:

```
To: rfc-info@ISI.EDU
Subject: getting fyis
help: ways_to_get_fyis
```

Also via e-mail, send a message containing a single word, `help`, to the address: **mailserv@ds.internic.net**. You may also use the World Wide Web clients (URL **ftp://ds.internic.net/rfc/** or **http://ds.internic.net**), Gopher (**ds.internic.net**), and FTP (**ds.internic.net**).

0001 F.Y.I. on F.Y.I.: Introduction to the F.Y.I. notes. G. S. Malkin, J. K. Reynolds. Mar-01-1990.

0002 FYI on a network management tool catalog: Tools for monitoring and debugging TCP/IP nets and interconnected devices. R. Enger, J. K. Reynolds. June 1993.

0003 FYI on where to start: A bibliography of Internetworking information. K. L. Bowers, T. L. LaQuey, J. K. Reynolds, K. Roubicek, M. K. Stahl, A. Yuan. Aug-01-1991.

0004 FYI on Questions and Answers-Answers to commonly asked "New Internet User" Questions. A. N. Marine, J. K. Reynolds, G. S. Malkin. March 1994.

0005 Choosing a name for your computer. D. Libes. Aug-01-1991.

0006 FYI on the X window system. R. W. Scheifler. Jan-01-1991.

0007 FYI on Questions and Answers: Answers to commonly asked "experienced Internet user" questions. G. S. Malkin, A. N. Marine, J. K. Reynolds. Feb-01-1991.

0008 Site Security Handbook. J. P. Holbrook, J. K. Reynolds. Jul-01-1991.

0009 Who's Who in the Internet: Biographies of IAB, IESG, and IRSG Members. G. S. Malkin. May 1992.

0010 There's Gold in Them thar Networks! or Searching for Treasure in All the Wrong Places. J. Martin. January 1993.

0011 A Revised Catalog of Available X.500 Implementations. A. Getchell, S. Sataluri, Editors. May 1994.

0012 Building a Network Information Services Infrastructure. D. Sitzler, P. Smith, A. N. Marine. February 1992.

0013 Executive Introduction to Directory Services Using the X.500 Protocol. C. Weider, J. K. Reynolds. March 1992.

0014 Technical Overview of Directory Services Using the X.500 Protocol. C. Weider, J. K. Reynolds, S. Heker. March 1992.

0015 Privacy and Accuracy Issues in Network Information Center Databases. J. Curran, A. N. Marine. August 1992.

0016 Connecting to the Internet-What Connecting Institutions Should Anticipate. ACM SIGUCCS. August 1992.

0017 The Tao of IETF-A Guide for New Attendees of the Internet Engineering Task Force. The IETF Secretariat, G. S. Malkin. November 1994.

0018 Internet Users' Glossary. G. S. Malkin, T. L. LaQuey Parker. January 1993.

0019 FYI on Introducing the Internet-A Short Bibliography of Introductory Internetworking Readings. E. Hoffman, L. Jackson. May 1993.

0020 FYI on What is the Internet?. E. Krol, E. Hoffman. May 1993.

0021 A Survey of Advanced Usages of X.500. C. Weider, R. Wright. July 1993.

0022 FYI on Questions and Answers-Answers to Commonly Asked "Primary and Secondary School Internet User" Questions. J. Sellers. February 1994.

0023 Guide to Network Resource Tool. EARN Staff. March 1994.

0024 How to Use Anonymous FTP. P. Deutsch, A. Emtage, A. N. Marine. May 1994.

0025 A Status Report on Networked Information Retrieval: Tools and Groups. J. Foster. August 1994.

0026 K–12 Internetworking Guidelines. J. Gargano, D. Wasley. November 1994.

0027 Tools for DNS debugging. A. Romao. November 1994.

Appendix 2: Excerpt of a Press Release from February 14, 1994

Two major developments toward the information superhighway received green lights Friday from the National Science Board (NSB) ... [T]he board gave its administrative approval for the National Science Foundation (NSF) to proceed with arrangements for a new architecture for the NSFNET computer network services, including an experimental expressway—called a very high speed Backbone Network Service (vBNS)....

The NSB ... approved a proposal that the agency's networking division award the operation of the vBNS to MCI Communications Corp. of Washington, DC.... The NSB also approved a plan by the networking division to establish two 5-year collaborative cooperative agreements to offer routing arbiter (RA) services for NSFNET and the Internet. Merit Network, Inc., a consortium of Michigan universities, will ... provide overall RA management and take the lead responsibility for a registry database, an operations center, and software development. The University of Southern California's Information Sciences Institute, working with IBM, will ... develop route servers, advanced routing techniques, a test bed, and routing engineering.

The vBNS will be a physically separate network to push the parameters of high-speed networking and to connect NSF's supercomputer centers, without affecting general network services.

The routing arbiter will be one of three major components of the new, advanced "general-use" network services. The other two components are projects calling for network access point managers and for regional network providers. With transmission capability of 45 million bits of data per second, the current NSFNET backbone makes the connections over commercially leased lines by linking 19 sites—called nodes—throughout the United States.

Developed by NSF in 1985, the NSFNET program long ago outgrew its initial vision: to provide broadband access to NSF's five supercomputing centers for researchers ... The new architecture calls for the regional networks to secure backbone services from commercial network providers, with a 5-year phasing out of NSF's backbone funding for the regional nets.

NSFNET Program Director Priscilla Huston said, "Our first priority as we make this transition is to ensure that the new infrastructure is robust and dependable before we shut down [currently scheduled for the end of April 1995] what has become a legendary national backbone service among the networking community. The new architecture will allow the research and education communities to take advantage of the excellent services available from commercial suppliers while it provides the interconnectivity that's essential for a vibrant national infrastructure.

"In 1987 you couldn't buy high-bandwidth Internet services. There were neither suppliers nor market. So we commissioned a high-bandwidth backbone; and, in the 7 years since, a number of comparable commercial services have arisen—catalyzed in part by the market generated by the NSFNET backbone. So now we can give backbone funding to the regionals and let them buy that connectivity from the suppliers of their choice."

chapter two

Electronic Mail

Jan K. Labanowski, Charles F. Bender, and Alejandro Pisanty

In This Chapter

This chapter discusses the mechanics of electronic mail (e-mail), the different types of e-mail addresses, and ways to find e-mail addresses.

3223–7/96/0025$13.00/1

WE ARE WITNESSING A REVOLUTION in computer communications. As with previous revolutions, some chaos is unavoidable. Standards for transmission of various kinds of information are only now beginning to emerge, and the tools are still difficult to configure and use. However, progress is being made. Computers are becoming as popular as televisions, and the original mission of the computer, that is, numeric calculations, is giving way to graphics, text processing, information retrieval, communication, visual conferencing, and entertainment. The computational power of inexpensive desktop computers is now substantially larger than that of supercomputers from only 20 years ago. The personal computers of today are equipped with powerful graphics capabilities, and the dumb, text-only terminals are dying fast.

Most important, computers can now communicate with each other, as opposed to the old days when computers were accessible only on-site. Computer networks today span the world and are now available in more than 100 countries. Access to computer networks is available not only to large organizations, but to homes and small businesses as well. And the most nagging problem of connecting to the network backbone, often referred to as "the last mile problem", is being solved by the progress in digital phone technology (faster modems and the growing popularity of the Integrated Services Digital Network or ISDN) and the deployment of affordable high-bandwidth cable television and communal fiber-optics networks.

Direct access to the Internet is very popular in academia, but it is still not generally available in homes or smaller organizations. The most obvious reason is the cost of such a service. On the other hand, electronic mail (or e-mail) is one of the least expensive Internet services (frequently available for free), as it is quite inexpensive to spool mail on a central computer and transmit and receive it during off-peak hours.

The unavailability of direct Internet access to and from all machines inside an organization is frequently dictated by security and other organizational reasons. If the intracompany computers contain sensitive information, preventing unauthorized access creates a substantial system administration overhead. Moreover, the intracompany network may not be using the TCP/IP (transmission control protocol/Internet protocol) required for Internet access.

Implementing a world-wide e-mail connection is relatively easy to do; all that is required is an Internet-accessible e-mail gateway that can also handle protocol conversion. Even this requirement can be circumvented by subscribing to an e-mail service offered by one of many Internet access providers. The simplicity of e-mail and the growing sophistication of e-mail readers ensures that it will be an integral part of electronic communications for many years to come.

This chapter will discuss e-mail from the UNIX point of view. There are many different approaches to e-mail; however, the basic principles are similar. In fact most of the labor involved in composing and delivering e-mail is performed by software, and for the end user the main difference will be the user interface. Many tools and utilities (both commercial, as well as free) provide automatic conversion between different e-mail protocols, formats, and encodings.

How Does e-mail Work?

As with traditional mail, sending electronic mail consists of several stages. Before mail is sent, it has to be composed, filed, addressed, and delivered to the post office (or retrieved from your post office box or mailbox and read and filed). Other operations are done at the post office, where mail is sorted, transported to the central distribution sites, and finally to a local post office and the addressee. Conceptually, e-mail is no different. In most cases two different programs are used: one for composing or reading the mail and another for sending, transporting, and collecting the mail.

Mail-User and Mail-Transfer Agents

Mail-user agents (MUAs) help you prepare, respond, forward, and dispose of your e-mail. Some of them are very advanced programs with easy-to-use graphical user interfaces, support for composite documents (i.e., documents that contain images, sound clips, or video together with the text), and assistance in filing and cataloging your messages. Some MUAs can pass mail within your local network, but delivering and receiving mail to and from a remote computer is done by the mail-transfer agent (MTA). In fact, you can consider your MUA as an interface to the MTA. The MTA is the actual program that delivers your mail to the addressee (or a relay machine if direct connection is not possible) and performs necessary conversions between different e-mail protocols. The MUA program provides tools for preparing or reading your own mail, and MTA serves as the postal service.

The operations of the MUA are important when you sit in front of your computer or terminal, but the responsibilities of the MTA continue after you log out. In general, when you send an e-mail message, the MUA passes the composed message to the MTA software. When you receive the message, the MTA signals your MUA that the new message is available. If you are logged out (or when your MUA software is not running) the MTA collects incoming messages in your system mailbox until your MUA looks at them.

Parts of an Electronic Message

The responsibility of the MUA is to compose the e-mail message so it conforms to the standards. The MUA also helps you in reading and processing your message. The message that you compose needs to start with important information. For this reason, each e-mail message consists of two parts: the header and the body (or text) of your message. Here is a sample message:

```
From apisan@redvax1.unam.mx Thu Sep 29 11:57:50 1994
Received: from redvax1.dgsca.unam.mx. for
 apisan@redvax1.dgsca.unam.mx
 by infomeister.osc.edu (8.6.9/930601.1506) id
 LAA19828; Thu, 29 Sep 1994 11:57:49 -0400
Received: from localhost (apisan@localhost) by
 redvax1.dgsca.unam.mx. (8.6.4/8.6.4) id JAA09292;
 Thu, 29 Sep 1994 09:56:33 -0600
Date: Thu, 29 Sep 1994 09:56:32 -0600 (CST)
From: Pisanty Baruch Alejandro-FQ
 <apisan@redvax1.dgsca.unam.mx>
To: Jan Labanowski <jkl@osc.edu>
Subject: How is the chapter progressing?
MUA'sIn-Reply-To:
 <199409261442.KAA14031@infomeister.osc.edu>
Message-ID: <Pine.SUN.3.90.940929095451.8131D-
 100000@redvax1.dgsca.unam.mx>
MIME-Version: 1.0
Content-Type: TEXT/PLAIN; charset=US-ASCII

Jan,
I hope that you are done with your portion of the
chapter. Send me your version and I will make the
corrections.
Alejandro
```

The header is separated from the body of the message by an empty line. In many cases the header may be even longer than the body, as in the example. Display of many of these header lines can be suppressed by configuring your MUA software, but they are important, and both MUAs and MTAs carefully examine the header to learn about the source of the incoming message and to treat the body of the message according to its contents. The header also helps in tracking delivery problems and in detecting phony e-mail messages.

The lines (fields) in the header start with the keyword, followed by a colon, a space, and the contents, which may sometimes occupy more than one line. The continuation lines start from a space or a tab character. Some fields in the header are obligatory; others are optional. Most of these fields are defined by the Internet computer documents called RFCs (requests for comments).[1]

Some fields are nonstandard and highly discouraged but used anyhow. The RFCs are constantly amended as new developments take place. The header fields and keywords are described in RFC-822[2]; however, numerous additions and extensions have been made to the originally proposed list of header keywords. Most notably, the MIME (multipurpose Internet mail extensions) that were introduced with RFC-1521[3] and RFC-1522[4] systematize, explain, and standardize ways in which e-mail can be used to send information other than plain text.[5] Also, approval for many proposed standards is pending in a form of pre-RFC. You can usually find them in the same archives that carry the RFCs. These texts frequently describe current common practice and have a high probability of being adopted as standards.

The message header is mostly prepared by the MUA program; however, some header fields are always added by the MTA (e.g., `Received:` lines). The MTA may also add essential fields if they were missing. The most important fields are described in Appendix 1.

Delivering the Message

Once the header and the body of your message are prepared by the MUA, the message is ready to be sent out. If your message is addressed only to the local users of your machine, the MUA usually performs direct delivery without the help of MTA (at least, this is how the popular UNIX *mail* program works). When at least one of the recipients is not local, the MUA passes your message to the MTA. The most popular MTA program on UNIX machines is called *sendmail*.[6] This program is not meant to be used by people who are not experienced users or network hackers. Being an MTA, *sendmail* has the difficult task of delivering your message. The remote machine may not be directly accessible from your site, and *sendmail* needs to find a gateway. When the remote machine is temporarily unavailable, *sendmail* has to batch your message and attempt delivery later without your intervention.

The MTA sometimes needs to alter or re-encode your message. The MTA must also ensure that the whole message is relayed toward the remote machine. At the same time, *sendmail* checks constantly to see if some remote machine wants to pass a message to local users. For these reasons, installation and customization of the MTA software is one of the more difficult tasks of system administration.

A popular protocol used for passing the message to a remote user is called simple mail-transfer protocol (SMTP), and is described in RFC-821.[7] Its extended version, ESMTP, is also available.[8] The simplicity and robustness of SMTP made it so attractive that its variants and analogs are used for many other tasks of intercomputer communications. In the Internet TCP/IP environment, TCP port number 25 is reserved for passing electronic mail. A possible transaction could look like this:

```
Connecting to postbox.acs.ohio-state.edu. (tcp)...
220 postbox.acs.ohio-state.edu Sendmail 8.6.9/5.901231
>>> HELO infomeister.osc.edu
250 postbox.acs.ohio-state.edu Hello infomeister.osc.edu
>>> MAIL From:<jkl@osc.edu>
250 <jkl@osc.edu>... Sender ok
>>> RCPT To:<jlabanow@magnus.acs.ohio-state.edu>
250 <jlabanow@magnus.acs.ohio-state.edu>... Recipient ok
>>> DATA
354 Enter mail, end with "." on a line by itself
>>>.
250 AAA05624 Message accepted for delivery
jlabanow@magnus.acs.ohio-state.edu... Sent
Closing connection to postbox.acs.ohio-state.edu.
>>> QUIT
```

In this log, the commands of the local machine (sender) are marked with >>> and consist of four-letter words. The responses of the remote machine (receiver) start with a number code. If the responses are multiline, the number is followed by a dash character (except on the last line. The contents after the number are usually ignored by the sender; however, ESMTP adds extensions in which the word following the number is relevant. The comments that follow the obligatory responses of the receiver are not important, and system administrators frequently change them to suit their senses of humor.

As the first step, the sender machine (in this case **infomeister.osc.edu**) must contact the remote machine (here **postbox.acs.ohio-state.edu**) and check to see if connection is possible. The remote machine is not always the machine of the intended recipient. The database of addresses has to be consulted. The address of the machine is supplied by the MX (mail exchange) record or the alias in the DNS (domain name system) database[9] (although a few years ago many installations still used the flat file named `hosts` that served as a database of known Internet addresses and had to be updated periodically). The DNS translates host names into numerical Internet addresses. This database is located on machines called "name servers" whose addresses are given to *`sendmail`* in its configuration file.

Address resolution may take several contacts with different machines, because the database is hierarchical and is distributed among many machines. The addresses local to each domain are stored and updated locally on their domain name server without a need for some central address depository. When you look at the machine name in the e-mail address, you see several portions separated by dots. In principle, the lookup starts from the rightmost portion and goes to the left side (however, one country adopted the reverse lookup direction).

The contact is made with the TCP port 25 of the remote machine. If there is no answer, the sender will abandon the contact and try again later. But in the example shown earlier the receiver was ready and answered with the code 220, which means "I am ready to play". At this point the sender introduced itself using the `HELO` command and gave its identity. The receiver should check if the calling machine is indeed the **infomeister.osc.edu**; if not, then it should log the real sender address (the real address is available in the IP packet requesting communication with port 25 of the receiver) as a precaution against unauthorized access. The receiver responded with a code 250, which states that it accepted the connection (although it could also say NO!!! if the machine is on the "enemy" list).

After getting an OK from the receiver, the sender presented its envelope address by issuing the command `MAIL From:`. This command gives the address of the account from which the mail is actually being sent. This address may be different from the one that appears on the `From:` line of the header, although they are usually the same. However, in mailing lists, the header `From:` address refers to the author of the message, and the envelope address is the address of the list administrator. It is done this way to direct possible errors to the administrator (or the computer program that administers the list), rather than the author, who usually does not even know who receives the message.

After verifying the envelope address, the receiver confirmed it with the code 250. Then, the sender provided the recipient address with the `RCPT To:` command. This is recipient's envelope address. There may be many recipients of the same message on `To:`, `Cc:`, `Bcc:`, etc. lines, and for each of them a separate envelope address has to be given. In a given session, only those recipients who will have their mail delivered via the current receiver machine are specified. For a message that contains recipients accessible via different machines, a separate connection with each appropriate machine has to be made for each recipient. Then, once the receiver learns who is the intended recipient, it needs to check if the mail can be delivered. If there is no such user on the receiver machine, or the user cannot receive mail (e.g., user's mailbox is full), the receiver answers with an error code (codes larger than 400). But in our case, the recipient was found `OK`, which was signaled to the sender by the code 250.

Now the sender will pass the actual message (with the header and the body). It signals it to the receiver with a command `DATA`, and after getting the "go ahead" via the code 345, it pumps the message to the receiver. By convention, the end of the message is signaled by a line that contains only a period. With some ancient software packages, you cannot have a line with a single period in your message; however, all modern packages check to see if there is a period as a first character of the line, and if there is, they add another one. So there are always either none or two or more periods at the beginning of each line in a valid message. The receiver, on the other hand, always removes the first period on the line, and if there is only a single period, it knows that the end of the message has been reached. After the receiver finds a single period on a line by itself, it responds with a code 250, saying that the message was received. At this point, the sender says "Good bye" by issuing the `QUIT` command (although in principle, it could start the transaction for another message). They disconnect afterwards (the receiver says "Good bye" by returning the code 221).

If at any stage of this dialog, one of the partners "dies", the message is considered not delivered, and the message stays in the mail queue on the sender machine. If the message cannot be successfully delivered within a few days (the timeout is usually set to 3 days), the message is returned to the author with a comment identifying problems in delivery.

Other Sender Commands

Besides the HELO, MAIL, RCPT, and DATA commands, some MTAs implement other commands. One of them is HELP, which frequently produces some short information about SMTP commands. The RSET command aborts the message in progress. Some optional commands, when implemented, allow manual testing for the particular recipient on a given machine. The VRFY command allows checking to see if the user or user's alias exists on a given machine. When the user is unknown, the receiver responds with an error code. The command EXPN expands aliases into real addresses. It is especially useful when the alias is a mailing list or the mail for a given user is being forwarded to another machine or alias. The use of EXPN and VRFY can be seen in the following example. Words in bold italic typeface (***user-input***) indicate input typed by the user, and words in plain typewriter typeface (screen output) indicate screen output from the computer. % is the system prompt.

```
% telnet infomeister.osc.edu 25
Connected to infomeister.osc.edu.
220-infomeister.osc.edu Sendmail
220 ESMTP spoken here
VRFY jkl
250 Jan Labanowski <jkl@infomeister.osc.edu>
VRFY bubulubu
550 bubulubu... User unknown
EXPN oscc
250-David Heisterberg <djh@infomeister.osc.edu>
250-Jan Labanowski <jkl@infomeister.osc.edu>
250 <jkl@ohstpy.mps.ohio-state.edu>
EXPN chemistry
250-Computational Chemistry <ccl@infomeister.osc.edu>
250 <"|exec /chem/forwarder.chem"@infomeister.osc.edu>
EXPN bumba
550 bumba... User unknown
VRFY trollius
250 <trollius@infomeister.osc.edu>
EXPN trollius
250 <trollius@tbag.osc.edu>
QUIT
221 infomeister.osc.edu closing connection
```

In this example, the alias oscc consisted of three addresses. The second EXPN example shows that mail sent to the chemistry alias will be forwarded to two places: a ccl user and the program *forwarder.chem*. In this case, the list of subscribers is not available because their addresses are known only to the program *for-*

`warder.chem`. A popular setup for larger lists allows the incoming message to be sent to a program rather than directly to the recipients. Such a program tests the authenticity of a message and may change its header to reflect bulk distribution. It can also send it for review or save messages automatically in the list archives.

The `VRFY` and `EXPN` commands are frequently disabled on a machine for security reasons. Knowing a user id makes the job of network crackers easier, and some organizations prefer to protect the identity of their staffs. Sometimes they are disabled in a tricky way; for example, they return the `2XX` code (valid address) for any name, but check the address only when a message is actually sent.

Different Types of e-mail Addresses

Most of us are used to addresses in the form of **somebody@some.domain**, but many different forms of addresses are in use. Several good books discuss the complexities of e-mail addresses (e.g., refs. 10 and 11). The most common form consists of two parts joined by the @ character: **user@domain.name.country**. The letter case does not matter, and the address: **user@domain.name.country** is equivalent to **uSEr@DomaiN.NAME.Country**. You can mix uppercase and lowercase letters freely, for example, to enhance readability, but traditionally Bitnet addresses are written in uppercase and UNIX addresses in lowercase. For historical reasons many addresses usually do not have the .us country code appended; they end in one of the following:

- **.com** for commercial institutions
- **.edu** for some Canadian and U.S. educational institutions
- **.gov** for U.S. government addresses
- **.mil** for U.S. military
- **.net** for networking organizations
- **.org** for nonprofit U.S. organizations
- **.int** for international organizations
- **.nato** for addresses within the NATO organization

If present, the country codes are two letters long; they are listed in Appendix 2. This appendix contains some countries that no longer exist politically, but some old addresses are still usable. For example, the .su (Soviet Union) domain is still working fine for most addresses. Many of the countries in Appendix 2 still do not have computer networks and e-mail, but in time they will.

How To Find the Current Status of e-mail in a Given Country

Retrieve the file `country-codes` from the directory `pub/usenet-by-group/news.answers/mail` from the anonymous FTP site **rtfm.mit.edu**, as follows. Words in bold italic typeface (***user-input***) indicate input typed by the user, words in plain typewriter typeface (`screen output`) indicate screen output from the computer, and comments are in regular italic in parentheses. % is the system prompt.

```
% ftp rtfm.mit.edu
Connected to rtfm.mit.edu.
220 rtfm ftpd; bugs to ftp-bugs@rtfm.mit.edu
Name (rtfm.mit.edu:ccl): anonymous
331 Guest login ok, give e-mail address as password.
Password: jkl@osc.edu (enter your e-mail address here)
230 Guest login ok, access restrictions apply.
ftp> cd pub/usenet-by-group/news.answers/mail
ftp> get country-codes
200 PORT command successful.
150 Opening ASCII mode data connection for country-codes
(31669 bytes).
226 Transfer complete.
local: country-codes remote: country-codes
32282 bytes received in 0.27 seconds
ftp> quit
221 Goodbye.
```

This file can also be retrieved via e-mail, as described later in this chapter.

Some countries (e.g., the United Kingdom and Japan) provide abbreviations: ac for academic institutions and co for companies. For example: **some-user@aaa.kyoto-u.ac.jp** is the address of some person at the Kyoto University, and the address: **some-user@rs6.big-comp.co.uk** may be an address of somebody working for a Big Company in the United Kingdom. The United Kingdom has two country abbreviations: .uk, which you will see most often; and .gb, which is used for X.500 protocol addressing. (X.500 addresses are discussed later.) Moreover, the addresses from the United Kingdom come sometimes in the reverse order, as **some-user@uk.co.big-comp.rs6**. Without entering into a discussion on who is right, you need to reverse the order of entries in the domain name to be **some-user@rs6.big-comp.co.uk**.

In these examples the @ character separates the user name (or alias) from the machine name. But sometimes the machine may not be directly accessible from your computer and you will need to specify a gateway that can contact the target machine. You may try to use a major machine in the proximity of your addressee as a gateway, for example, some machine at the local university in the town or country.

How To Specify a Gateway

Replace the @ in the simple address with a % character and append the address of the gateway: **user%some.machine@the.gateway**

For example, your machine may not have direct access to Bitnet. In this case, the address: **user@node.bitnet** needs to be converted **user%node@some.gateway** or **user%node.bitnet@some.gateway**. Some machines, called Bitnet gateways, are accessible both via Bitnet and via Internet (in fact most Bitnet machines use the Internet TCP/IP protocol to communicate). The users on such machines have two addresses, the Bitnet one (e.g., **JKL@OHSTPY.BITNET**, although the .BITNET part is frequently not included and the address may be written as **JKL@OHSTPY** or even **JKL AT OHSTPY**) and the corresponding Internet one (e.g., **jkl@ohstpy.mps.ohio-state.edu**). If you do not have access to a Bitnet gateway, and your machine does not forward your message automatically to a Bitnet node, you can still send your message to a Bitnet address by using a Bitnet gateway. For best results use a Bitnet gateway that is close to the recipient. If everything else fails, you can use the official Bitnet/Internet gateway: **cunyvm.cuny.edu**, (e.g., send your mail to the address modified to **jkl%ohstpy@cunyvm.cuny.edu**).

Other address forms are in use. For example, CompuServe addresses use a different form.

How To Send Mail to a CompuServe Address

Use the form **12345.6789@compuserve.com** (the digits represent the CompuServe account number, here 12345,6789).

There are also old-style addresses, the UUCP (UNIX-to-UNIX-copy) paths, coming from the times when computers exchanged information usually via modems over phone lines. In those days, each machine had its own name (restricted to eight characters), and it knew only about a few other machines for which it stored their phone numbers. To send e-mail from one machine to another, not only the name of the target machine had to be known, but also the complete path. This path provided a series of machine names, separated by exclamation points (called "bangs") in which the previous machine knew how to contact the next one. For example, the address **host1!host2!host3!john** means that the message needs to go first to the machine named host1, where it will be relayed to host2, which knows how to contact host3, where john has an account. Luckily, such addresses are becoming extinct. If you ever see one, ask your system administrator how to convert it to the Internet-style address. You may also try to change it to the form **john%host3@some.machine** (e.g., **john%host3@uunet.uu.net**) if you know who provides the UUCP connection for the last host on the path.

The other type of address, X.500 address,[12,13] is becoming more and more popular. It originated from the Open Systems Interconnect (OSI) protocol. The advantage of X.500 is that it is hierarchical and breaks the address into portions describing the country, district, city, organization, department, and name. This system makes it an easy object to be searched with a computer. However, it is long, convoluted, and hard to memorize or type. It is used by some commercial e-mail providers (e.g., Sprint) and consists of fields separated by slashes, for example, PN=JOHN.M.SMITH/DD.ID=SSW.JMSMITH/PRMD=BIGCOM/ADD=TELEMAIL/C=US/@ sprint.com. More details on e-mail addressing can be found in several books on this topic.[11,14,15]

Finding e-mail Addresses

Now, suppose you do not know the e-mail address of somebody and want to find it out. The easiest way is to call this person on the phone. Several resources can help you find e-mail addresses, but there is no guarantee that you will be successful. Even if you can find the address, it may be one that is no longer in use. Excellent guides on finding addresses are available by anonymous FTP from **rtfm.mit.edu**.

How To Retrieve Electronic Guides to e-mail Addresses

These guides can be retrieved via e-mail by sending the following message:

```
send usenet-by-group/news.answers/mail/college-email
send usenet-by-hierarchy/comp/answers/finding-addresses
```

to **mail-server@rtfm.mit.edu**. In this case, the files requested are directory names, and the server will send you a list of files in response. The guides usually come in several parts, and you need to send a message again, this time requesting the actual files:

```
send usenet-by-group/news.answers/mail/college-email/part1
send usenet-by-group/news.answers/mail/college-email/part2
send usenet-by-group/news.answers/mail/college-email/part3
```

How To Find an Address Using the Netfind Service

The most popular address-finding method is the Netfind service. Telnet to the Netfind service that is closest to the location of your addressee. Some popular Netfind sites are

archie.au
bruno.cs.colorado.edu
dino.conicit.ve
krnic.net
ds.internic.net
lincoln.technet.sg
malloco.ing.puc.cl
monolith.cc.ic.ac.uk
mudhoney.micro.umn.edu
netfind.if.usp.br
netfind.anu.edu.au
netfind.oc.com
netfind.sjsu.edu
nic.nm.kr
netfind.vslib.cz
nic.uakom.sk
redmont.cis.uab.edu

Words in bold italic typeface (***user-input***) indicate input typed by the user, and words in plain typewriter typeface (`screen output`) indicate screen output from the computer. % is the system prompt.

```
% telnet bruno.cs.colorado.edu
Trying 128.138.243.150 ...
Connected to bruno.cs.colorado.edu.
login: netfind
```

Then follow the instructions on the screen. Choose 2 (search) from the menu. Give the person's name or login id as a first item, and then enter locators, (e.g., `labanowski supercomputer columbus`, or if you know the login id and/or domain, `jkl osc`). Then choose promising machines from the list, and follow the prompts.

How To Find an Address Using the `whois` Command

The other popular service is the `whois` command of UNIX, but you need to know the `whois` server that may contain the person under question, for example, `whois -h osu.edu labanowski`.

If you do not have a `whois` program, you may still try `telnet osu.edu 43` and then enter the name on the next line. The list of available `whois` servers can be retrieved from the anonymous FTP/mail-server **rtfm.mit.edu** in the directory `pub/whois`. You can also retrieve this information either by using your gopher client software and typing ***gopher sipb.mit.edu***, or you can connect to it by opening the **gopher://sipb.mit.edu** URL (uniform resource locator) in your Web browser and pointing to `Internet whois servers` on the menu. This gopher site actually allows you to search the various `whois` servers. You may also find a list of people who recently posted something to the Usenet newsgroups in the directory `pub/usenet-addresses/lists` of the anonymous ftp server **rtfm.mit.edu**.

If you know that the person is active on Internet, it is an excellent resource for obtaining recent e-mail addresses. The Knowbot Information Service offers a search of several databases. For example, to look for Charlie Bender you could proceed in the following way:

```
% telnet info.cnri.reston.va.us 185
> help
... a long stream of information follows ...
> country us
> query bender
... a long list of Benders follows ...
> quit
```

Again, words in bold italic typeface (***user-input***) indicate input typed by the user, words in plain typewriter typeface (`screen output`) indicate screen output from the computer, and words in regular italic are comments. % is the system prompt.

Besides using specialized sites and software to look for the e-mail address, you may want to find a gopher for the institution where this person works. (This route is especially helpful for people working in academia.) It is becoming more and more popular for larger organizations to provide phone and address directories of their staff via gopher search.

Conclusions

Space limitations do not allow us to provide many other important uses of e-mail. But with its growing popularity and low cost, e-mail functionality is driven by the community needs. An e-mail interface to Gopher allows searching and retrieving Gopher items even if you do not have direct Internet connection.

How To Search and Retrieve Gopher Items via e-mail

Send a single line, `help`, to **gophermail@calvin.edu** and follow the instructions.

World Wide Web access via e-mail is also being tested.

Similarly, you can retrieve files from various anonymous FTP sites by using *`ftpmail`*. Learn more about it by sending a one-line message (`help`) to one of the many *`ftpmail`* providers:

ftpmail@sunsite.unc.edu
ftpmail@decwrl.dec.com
ftpmail@pucc.princeton.edu
ftpmail@grasp.insa-lyon.fr
ftpmail@lth.se
ftpmail@ftp.uni-stuttgart.de
ftpmail@doc.ic.ac.uk

The syntax of commands is very similar to the FTP program commands: You open the connection with the `open some.ftp.site.address` command and use the `cd`, `get`, and `dir` commands. For big files you need to set the chunk size with the `size 60k` command. You can also specify the address to which the files should be sent by using `reply-to some.other.address` command when *ftpmail* cannot retrieve your correct address from the header of your message.

The Archie servers can also be queried via e-mail. Archie is a service that helps you locate files and programs on the thousands of registered anonymous FTP sites. You can get instructions on using Archie by sending a single line, `help`, to one of the servers:

archie@archie.ans.net
archie@archie.rutgers.edu
archie@archie.suran.net
archie@archie.mcgill.ca
archie@archie.funet.fi
archie@archie.au
archie@archie.au
archie@archie.doc.ic.ac.uk
archie@archie.wide.ad.jp
archie@archie.ncu.edu.tw.

For example, if you are looking for a program called *mopac*, send the following message:

```
set search sub
find mopac
quit
```

to one of the Archie servers and you will receive a list of anonymous FTP sites where this program (or a file that has *mopac* in its name) is stored.

You can use e-mail to participate in Usenet discussions via many mail-to-news interfaces. This information can be retrieved from the anonymous FTP server **rtfm.mit.edu** from the directory `/pub/usenet-by-group/news.answers/mail/news-gateways`. You can also get these files via e-mail by using the mail server for this site (**mail-server@rtfm.mit.edu**) as described earlier in this chapter.

So, you may ask, "Why do I need full Internet access if everything can be done via e-mail?" The answer is that you need it because the access given to these services via e-mail is noninteractive, that is, you will have to wait for the answer, sometimes as long as a day (if your mail is relayed or batched on some machine to save cost). Although e-mail is a very versatile and powerful tool, it is best used for performing the task for which it was created: sending electronic mail.

References

[1]RFCs represent the latest status of the Internet protocols. Some are written in a highly technical language, but most of them provide easy-to-follow overviews. The RFCs discuss all aspects of Internet, not only e-mail. They are available at many computer archives all over the world. For example, they can be retrieved via WWW (**http://ds.internic.net**), Gopher (**ds.internic.net**), FTP (**ds.internic.net**), and e-mail (send a message containing a single word, `help`, to the address: **mailserv@ds.internic.net** for details on retrieval). RFCs are discussed in more detail in Chapter 1.

[2]Crocker, D. "Standard for the Format of ARPA Internet Text Messages", 1982, RFC-822.

[3]Borenstein, N.; Freed, N. "MIME (Multipurpose Internet Mail Extensions) Part One: Mechanisms for Specifying and Describing the Format of Internet Message Bodies", 1993, RFC-1521.

[4]Moore, K. "MIME (Multipurpose Internet Mail Extensions) Part Two: Message Header Extensions for Non-ASCII Text", 1993, RFC-1522.

[5]To find pointers to other applicable RFCs, retrieve the file `/rfc/rfc-index.txt` or search the `ds.internic.net` archives by keyword.

[6]Costales, B.; Allman E.; Rickert N. *sendmail;* O'Reilly & Associates: Sebastopol, CA, 1993.

[7]Postel, J. "Simple Mail Transfer Protocol", 1982, RFC-821.

[8]Klensin, J. Freed, N.; Rose, M.; Stefferud, E.; Crocker, D. "SMTP Service Extensions", 1993, RFC-1425.

[9]Liu, C.; Albitz, A. *DNS and BIND;* O'Reilly & Associates: Sebastopol, CA, 1992.

[10]Frey, D.; Adams, R. *A Directory of Electronic Mail, Addressing, and Networks;* O'Reilly & Associates: Sebastopol, CA, 1994.

[11]Krol, E. *The Whole Internet;* O'Reilly & Associates: Sebastopol, CA, 1994.

[12]Getchell, A.; Sataluri, S. "A Revised Catalog of Available X.500 Implementations", 1994, RFC-1632.

[13]Jurg, P. "Introduction to White Pages Services Based on X.500, 1994", RFC-1684.

[14]*The Internet Unleashed;* Kelly, K., Ed.; Sams Publishing: Indianapolis, IN, 1994.

[15]Shirky, C. *The Internet by e-mail;* Ziff-Davies Press: Emerville, CA, 1994.

Appendix 1: Message Header Fields

`From` (not followed by a colon) is typical for UNIX systems running a *`sendmail`* MTA program. It is the first line of the message. This line provides the envelope address of the sender, that is, the address that was given to the addressee machine when the two MTA programs (the sending MTA and the receiving MTA) were negotiating the message. It also contains the time and the date of the transaction.

`From:` specifies the address of the author of the message. This line frequently contains the full name of the person.

`Reply-To:` if given, is a preferred address to which the reply should be directed. It may be different from the contents of the `From:` line. It is used when the author sent a message from an account that is used infrequently or is unreliable (e.g., computer is powered down for the night). It may also be used to identify additional people to whom the reply should be directed (e.g., in a teleconference). This line is set by the originator of the message.

`Return-Path:` is rarely seen in messages. It may be present in relayed messages and is added by the final MTA. It provides a verified address of the originator when the generic address on the `From:` line needs to be augmented with the path in order for the reply to be successful. (In some cases, the % is needed for specifying the path for delivery).

`Sender:` may be different from the `From:` line, for example, in massive distributions like electronic lists or other broadcasts. It provides the address of the list administrator or some other agent, rather than the the address of the original author. This address should be used to report delivery errors or other problems and should not be used for replies. When the `Sender:` address is identical with the author address on the `From:` line, the `Sender:` field is not needed and should not be used.

`Errors-To:` provides an address (usually administrative) to which the messages about delivery problems should be forwarded.

`To:` gives one or more addresses of primary recipients. In mailing lists, this field may not have your own address, but the generic address of the list.

`Cc:` enumerates the addresses that received a carbon copy of the message.

`Bcc:` You will probably never see this field, though it exists when a message is passed from the MUA to the MTA. This line lists the blind carbon copy addresses, that is, the addresses that received the copy of the message but are not listed in the header (the MTA removes this header field before sending the message). This option is used when you want to send a copy of the message to some people, but you do not want their addresses to be noted in the message.

`Subject:` describes the message. This header field is not obligatory, but is highly recommended, because people use it to quickly scan their mail. Make it short and informative. Subjects like "e-mail message" and "information you requested" are useless.

`Apparently-To:` and `Apparently-From:` are added by the MTA when the original message does not have valid header lines `To:` or `From:`. The MTA inserts the envelope addresses, that is, the addresses that were presented to it when the actual mail transaction occurred according to the SMTP protocol.

`Received:` One or more such header lines (usually with continuation lines) logs the path that the message has traveled. Each gateway passed on the road between the originator and you leaves its name and the time that the message was relayed. Sometimes when a wrongly configured machine corrupts the address on the `From:` line, the `Received:` lines are the only way to establish the full address of the message originator. These lines can also be used to detect falsified messages. The more intelligent MTAs also check the lines for possible address loops or other delivery problems. The message is usually rejected when there are more than 30 `Received:` fields, because this large number of them indicates the possibility of a mail loop (e.g., somebody forwards the mail to the same e-mail address to which messages are actually coming).

`Message-Id:` is a unique tag that marks the message. It may be used to locate a particular message in the mail folder.

`Date:` provides the date and time that the message originated. It also gives the difference between Universal or Greenwich Mean Time (UT or GMT) and the local time, and may specify the local time zone (e.g., EST or EDT).

`In-Reply-To:` is a reference to the message that is being replied to. It usually contains the `Message-Id:`, but can optionally include the date and the author address of the previous message.

`Content-Length:` is becoming popular to specify the length of the original message. It is used by some MUAs to check the integrity of the message (i.e., to ensure that the message was not truncated).

`MIME-Version:` specifies that the message is prepared by a MUA that knows about MIMEs (multipurpose Internet mail extensions). The MIMEs add power to e-mail because they remove the restrictions on text (allowing for languages and alphabets other than U.S. English) and provide for images, audio files, videos, binary files, multipart messages, etc. This field does not provide any other information besides the MIME version number to ensure that the related fields are properly interpreted. RFC-1521[3] defines associated fields (`Content-Type:`, `Content-Transfer-Encoding:`, and `Content-ID:`) that provide information about the format, type, and contents of the message body. At present, the `Content-Type:` field is usually: text/plain; charset=US-ASCII; however, some programs already support MIME extensions (e.g., the *`Elm`* program by Dave Taylor, *`Pine`* e-mail from University of Washington, or *`metamail`* by N. Borenstein of Bellcore). The number of intelligent MUAs that can interpret MIME and present them to the mail recipient is constantly growing. The concept of MIME is a major, relatively recent, and long overdue development in the Internet e-mail arena. It was enthusiastically accepted by software developers. The speed with which it was incorporated in so many e-mail packages ensures its viability.

`Resent-From:` and other fields starting with `Resent-` are added by some systems to identify that the message was forwarded. The original fields `From:`, `To:`, `Cc:`, etc., are not changed and appear as they were in the original message. The `Resent-` fields are added by the forwarding system; that is, they are more recent.

`Return-Receipt-To:` is present when the sender requests a confirmation of delivery. The confirmations will be sent to the address given on this line. You should not request delivery confirmation when sending a message to an electronic list. You may end up with hundreds of confirmations in your mailbox and also flood the subscribers with confirmations of confirmations. Most modern electronic lists remove this field before sending the message to subscribers, but some are still not aware of its possible side effects.

`Keyword:` may be used to specify keywords for automatic text processing.

`Comments:` provides an option to enter descriptive text about the message. In most cases this field provides the name of the software used to compose or send the message.

`X-something` gives an opportunity for creating additional header fields. The header keyword starting from X is not processed by the standard versions of MUA and MTA software and will not interfere with new keywords in the future. These fields are frequently used to comment out original header fields (e.g., `X-Return-Receipt-To:`) when a message is modified or converted to another e-mail protocol. It is also used to introduce experimental header fields that are understood only by specially designed software.

Appendix 2: ISO-3166 Country Codes Used in Addressing Electronic Mail

.ad	Andorra
.ae	United Arab Emirates
.af	Afghanistan (Islamic St.)
.ag	Antigua and Barbuda
.ai	Anguilla
.al	Albania
.am	Armenia
.an	Netherland Antilles
.ao	Angola (Republic of)
.aq	Antarctica
.ar	Argentina
.as	American Samoa
.at	Austria
.au	Australia
.aw	Aruba
.az	Azerbaijan
.ba	Bosnia-Herzegovina
.bb	Barbados
.bd	Bangladesh
.be	Belgium
.bf	Burkina Faso
.bg	Bulgaria
.bh	Bahrain
.bi	Burundi
.bj	Benin
.bm	Bermuda
.bn	Brunei Darussalam
.bo	Bolivia
.br	Brazil
.bs	Bahamas

.bt	Bhutan
.bv	Bouvet Island
.bw	Botswana
.by	Belarus
.bz	Belize
.ca	Canada
.cc	Cocos (Keeling) Isl.
.cf	Central African Rep.
.cg	Congo
.ch	Switzerland
.ci	Ivory Coast
.ck	Cook Islands
.cl	Chile
.cm	Cameroon
.cn	China
.co	Colombia
.cr	Costa Rica
.cs	Czechoslovakia
.cu	Cuba
.cv	Cape Verde
.cx	Christmas Island
.cy	Cyprus
.cz	Czech Republic
.de	Germany
.dj	Djibouti
.dk	Denmark
.dm	Dominica
.do	Dominican Republic
.dz	Algeria
.ec	Ecuador
.ee	Estonia
.eg	Egypt
.eh	Western Sahara
.er	Eritrea
.es	Spain
.et	Ethiopia
.fi	Finland
.fj	Fiji
.fk	Falkland Isl.
.fm	Micronesia
.fo	Faroe Islands
.fr	France
.fx	France (European Ter.)
.ga	Gabon
.gb	Great Britain (UK)
.gd	Grenada
.ge	Georgia
.gf	Guyana (Fr.)
.gh	Ghana
.gi	Gibraltar
.gl	Greenland
.gm	Gambia
.gn	Guinea
.gp	Guadeloupe (Fr.)
.gq	Equatorial Guinea
.gr	Greece
.gs	S.Georgia S.Sandwich Isl.
.gt	Guatemala
.gu	Guam (US)
.gw	Guinea Bissau
.gy	Guyana
.hk	Hong Kong
.hm	Heard McDonald Isl.
.hn	Honduras
.hr	Croatia
.ht	Haiti
.hu	Hungary
.id	Indonesia
.ie	Ireland
.il	Israel
.in	India
.io	British Indian O.Terr.
.iq	Iraq
.ir	Iran
.is	Iceland
.it	Italy
.jm	Jamaica
.jo	Jordan
.jp	Japan
.ke	Kenya

.kg Kyrgyz Republic
.kh Cambodia
.ki Kiribati
.km Comoros
.kn St.Kitts NevisAnguilla
.kp Korea (North)
.kr Korea (South)
.kw Kuwait
.ky Cayman Islands
.kz Kazachstan
.la Laos
.lb Lebanon
.lc Saint Lucia
.li Liechtenstein
.lk Sri Lanka
.lr Liberia
.ls Lesotho
.lt Lithuania
.lu Luxembourg
.lv Latvia
.ly Libya
.ma Morocco
.mc Monaco
.md Moldova
.mg Madagascar (Republic of)
.mh Marshall Islands
.mk Macedonia (formerYugo.)
.ml Mali
.mm Myanmar
.mn Mongolia
.mo Macau
.mp Northern Mariana Isl.
.mq Martinique (Fr.)
.mr Mauritania
.ms Montserrat
.mt Malta
.mu Mauritius
.mv Maldives
.mw Malawi
.mx Mexico
.my Malaysia
.mz Mozambique
.na Namibia
.nc New Caledonia (Fr.)
.ne Niger
.nf Norfolk Island
.ng Nigeria
.ni Nicaragua
.nl Netherlands
.no Norway
.np Nepal
.nr Nauru
.nu Niue
.nz New Zealand
.om Oman
.pa Panama
.pe Peru
.pf Polynesia (Fr.)
.pg Papua New Guinea
.ph Philippines
.pk Pakistan
.pl Poland
.pm St. Pierre Miquelon
.pn Pitcairn
.pr Puerto Rico (US)
.pt Portugal
.pw Palau
.py Paraguay
.qa Qatar
.re Reunion (Fr.)
.ro Romania
.ru Russian Federation
.rw Rwanda
.sa Saudi Arabia
.sb Solomon Islands
.sc Seychelles
.sd Sudan
.se Sweden
.sg Singapore
.sh St. Helena

.si	Slovenia
.sj	Svalbard Jan Mayen
.sk	Slovakia (Slovak Rep)
.sl	Sierra Leone
.sm	San Marino
.sn	Senegal
.so	Somalia
.sr	Suriname
.st	St. Tome and Principe
.su	Soviet Union
.sv	El Salvador
.sy	Syria
.sz	Swaziland
.tc	Turks Caicos Islands
.td	Chad
.tf	French Southern Terr.
.tg	Togo
.th	Thailand
.tj	Tadjikistan
.tk	Tokelau
.tm	Turkmenistan
.tn	Tunisia
.to	Tonga
.tp	East Timor
.tr	Turkey
.tt	Trinidad Tobago
.tv	Tuvalu
.tw	Taiwan
.tz	Tanzania
.ua	Ukraine
.ug	Uganda
.uk	United Kingdom
.um	US Minor outlying Isl.
.us	United States
.uy	Uruguay
.uz	Uzbekistan
.va	Vatican City State
.vc	St.Vincent Grenadines
.ve	Venezuela
.vg	Virgin Islands (British)
.vi	Virgin Islands (US)
.vn	Vietnam
.vu	Vanuatu
.wf	Wallis Futuna Islands
.ws	Samoa
.ye	Yemen
.yt	Mayotte
.yu	Yugoslavia
.za	South Africa
.zm	Zambia
.zr	Zaire
.zw	Zimbabwe

NOTE: ISO is the International Standards Organization in Geneva, Switzerland.

chapter three

The Berkeley Mail Program

Jan K. Labanowski, Charles F. Bender, and Alejandro Pisanty

In This Chapter

This chapter gives instructions for using the Berkeley mail program on UNIX machines.

3223–7/96/0051$12.00/1

MANY PROGRAMS ALLOW YOU TO COMPOSE and read your electronic mail (e-mail). They differ in capabilities and may have different user interfaces, but most of them address the basic e-mail functions. The examples that follow assume that you are using the Berkeley mail program in UNIX to compose and read your mail. We chose to describe the Berkeley mail program because it is still the most popular mail-user agent (MUA; explained in detail in Chapter 2), and at the same time it is not particularly user-friendly. On the other hand, if you master Berkeley mail, you will not have any problems with many of the full-screen MUAs that are slowly replacing the original Berkeley workhorse.

Berkeley mail also has advantages if you read your mail using a slow modem because it is very sparing as far as transmission is concerned. On UNIX workstations it is usually located in the `/usr/ucb` directory. The full-screen X-windows version of this program is called *`xmail`*. On machines that run System V UNIX, the Berkeley mail program is called *`mailx`*, or *`Mail`*, and the *`mail`* program is a System V MUA with very limited capabilities.

Reading and Dealing with e-mail

To read your mail, simply type `mail` (or `mailx`, if you are on System V). The program will list the directory of your mail with `From:` addresses, dates and `Subject:` lines. Each message has a corresponding number, listed in the first column. After listing messages, the *`mail`* program enters command mode and displays the prompt (usually the & or ? character). By default, *`mail`* operates on your system mailbox. The system mailbox can be found in the `/usr/spool/mail` directory, and its name corresponds to the user login name. Messages are placed there by *`sendmail`* as they arrive. A new message is simply appended to the end of the system mailbox.

Each message starts with the line

```
From some.address date
```

Other MUAs may use a different convention to mark the start of a new message. You should avoid the word "From" as the first word on a line in your message. However, in most cases, it will be changed to ">From" automatically by your software to prevent it from being taken as the message start.

Besides the system mailbox, you also have a standard mailbox file that is usually called `mbox` and resides in your `HOME` directory (although you can place it elsewhere by editing a `.mailrc` file). The file `mbox` is a default folder in which the messages that you have read are saved.

Berkeley mail allows messages to be read, deleted, saved in a file or folder, forwarded, or processed (piped) through other programs.

- To print your message on your default printer, enter `| lpr` after the & prompt.
- To view a message, use the `p #` command. The # character will be used in the following descriptions to denote the message number (or a range or group of messages, e.g., 3–7 or 3,4,6,7). The `p` is called the print command, although it does not print anything, but simply displays the message on the terminal (this name comes from the times when teletypes were used as terminals).
- To save a message in a file or folder, type `s # somefile`, where # is the message number or numbers, as already defined. It is only natural that you will want to keep important messages. It is best to save them in mail folders, which are selected files residing in a separate directory. Choose easy-to-remember names for your folders. Also, the folder names should be related clearly to the messages that will be stored in them. You may want to decide in advance which e-mail folders you will need. If you create too many of them, you will have a hard time managing them, but, on the other hand, maintaining only one or two folders will make it difficult to find the message that you need among the many other messages.
- The `s` command appends the message to the end of the file or creates a new file if needed. If you prefix the file name with a +, (e.g., `s 3 +conferences`), the message (in this case number 3) will be saved in the folder directory (to be explained later) in the file `conferences`. The command `s 3 conferences` would save the message in the file `conferences` in your current directory.
- If you deleted a message with the `d #` command, you can undelete it before leaving mail by using the `u #` command.
- To see only the `From:` and `Subject:` lines of messages, enter the `f #` command.
- To reply to message #, enter the `r #` or `R #` command. In this and other commands, if the # is omitted, the current message is taken. The difference between the `R` and `r` commands is that `R` sends the reply to all the recipients of the message (i.e., also the ones on the

Cc: line), and r sends the reply only to the author of the message. Unfortunately, these may be set up differently on your system, and you may need to use the ? (help) command.

- To exit mail, enter x or q. The x command leaves your mailbox intact, but the q command may remove messages that were deleted with the d # command or saved in a file with the s # somefile command. Also, depending on the options you have selected, the q command may move the messages that you have already viewed from your system mailbox to your standard mailbox (usually mbox in your HOME directory).

Sending e-mail

You can send mail in two ways. If you are currently reading mail, the m recipient (for example: m jkl@osc.edu) command will switch you to send mode. If you are not currently inside the mail program, you can type mail somebody@some.machine and you will be prompted for the text of the message. When typing a message, you can also execute commands by placing the ~ character as the first character on the line and following it with a command:

- To read a file into your message use the ~r somefile command, where somefile is the specific file name.
- To set your Subject: line use the ~s some-text command.
- The ~e and ~v commands will call an editor, so you can modify the text already entered.
- The ~c address, ~b address, and ~t address commands add addresses to the Cc:, Bcc:, and To: fields in the header, respectively.
- The ~? command lists a short description of available commands.
- The ~w somefile command writes the currently typed message into a file, where somefile is the specific file name.
- The ~a and ~A commands allow you to append a predefined signature to your message, as described later.
- You can also use the read mode commands when composing a message by prefixing them with the ~: character sequence. For example, to read the last message in your mailbox, type the ~:p $ command (the $ is a shorthand for the last message in your mailbox).

- To incorporate or forward messages that you received previously, use the `~m #` or with the `~f #` commands (the `m` command will indent each line of the included message with a tab character, but `f` will not).
- The `~| some-command` will pass the composed message through some program.
- The `~!command` will allow you to execute a command or program from within *mail*.
- To end the message enter the EOF (end of file) character (in UNIX it is usually a Ctrl-D character) or enter the `~.` command.
- Finally, if you want to cancel the message, the `~q` command will abort sending it, and will save the currently composed message in the file called `dead.letter`. You can also abort sending a message with the `~x` command, but in this case all currently typed text will be lost.

Customizing Your Mail

For optimum performance you may need to tailor mail by editing the file `.mailrc` in your HOME directory.

It is popular to set up aliases for frequently used addresses. For example, the entry:

```
alias mystudents jeff@horse, betty@cow, dan@piglet
alias mystudents willy@labsgi, kris@sparc, trev@crayon
alias lori lbk%bmgvms.bitnet@ohstpy.mps.ohio-state.edu
```

will allow you to send messages to six addresses by sending mail to the group called "mystudents". The alias is also convenient for long or difficult to remember addresses. The `.mailrc` file allows you to choose the preferred editor called with the `~e` or `~v` command:

```
set EDITOR=vi
set VISUAL=xemacs
```

You may also specify two signatures to be used with the `~a` and the `~A` commands:

```
set sign="\nDick rfs@ucxx.edu\nLife is short \
sleep less.\n"
Set Sign="\nDr. Richard F. Smith\nDepartment \
of Chemiology\nUniversity of South Columbus\n\
Columbus, OH 43221\nEmail: rfs@ucxx.edu\n"
```

Your signature needs to be entered as a single line in `.mailrc` (sometimes a very, very long one, although it can be split by entering a \ immediately before the line break, as in the previous example). However, it does not have to be a single line in your message because you can include the new-line character (`\n`) in your string.

It is also useful to request that all mail that you are sending is collected in a file:

```
set record=/home/janl/folders/outmbox
```

This way, if you forget to include your name on the `Cc:` line, you will still have a copy of your message saved. The `.mailrc` file allows you to set or unset some variables that affect mail operation, e.g., the line:

```
set metoo askcc asksub nokeepsave nohold noautoprint
```

is interpreted in the following way:

`metoo` requests that mail sent by you to an alias that contains your address is delivered to you, that is, you get the copy of your own message. The `nometoo` option deletes your address in such a case, and you do not receive the copy of your own message. This is an issue if you send a message to a local mailing list to which you belong. The `metoo` will force a copy of your message to be sent to you. This message is no news to you, but it is nice to have a confirmation that it was distributed.

`asksub` will cause mail ask you for the `Subject:` line when composing the message.

`askcc` instructs mail to prompt you for the `Cc:` (carbon copy) addresses after you finish the message.

`nokeepsave` will make messages that you explicitly save in some file be automatically purged from your system mailbox. The keepsave will make these messages stay in your system mailbox until you explicitly delete them with a `d #` command.

`nohold` tells mail to move messages from your system mailbox to the standard mailbox after they have been read. The `hold` option would preserve messages in the system mailbox after you read them.

`noautoprint` will inhibit displaying the message after it has been deleted or undeleted. The autoprint option will force showing the messages being deleted or undeleted.

It is also convenient to specify the directory in which you will keep your mail folders, that is, the files in which you save selected messages for future reference. For example, if you want the folders to exist in the `mail-folders` subdirectory under your `HOME` directory, place the following line:

```
set folder=mail-folders
```

in the `.mailrc` file. To save messages in the folders in this directory, you need to prefix the folder name with a + character; for example, `s 2 +frank` will append the second message to the file `HOME/mail-folders/frank`, where `HOME` is the directory name of your home directory. This option is very convenient if you use a graphical version of mail, such as *`xmail`*, because you do not have to type the folder name but can choose it from a menu.

The UNIX *`mail`* program does not interpret the MIME-Version and associated header fields. (MIME is multipurpose Internet mail extension, discussed in detail in Chapter 6.) This major deficiency can be overcome by migrating to the more powerful e-mail software like *`Elm`* and *`Pine`*. The list of commercially and publicly available software supporting multipurpose extensions is growing fast. However, Berkeley mail is quite flexible, so it can be interfaced with the publicly available *`metamail`*[1] toolkit for processing MIME mail. The easiest way is to pipe a message through *`metamail`* when reading mail by typing `| metamail` (or `| metamail -w` if you did not create a `.mailcap` file).

The trickier way to interface with *`metamail`* is to set the PAGER variable in the `.mailrc` file. The PAGER variable is used to set the name of the program used for displaying a message, a page at a time. An associated variable, `crt`, can be set to a number of lines on a page, (e.g., 22). When a message is longer than the contents of `crt`, the *`mail`* program automatically pipes the message through the pager program specified by the PAGER variable (by default, the `more` pager is used). You can fool *`mail`* with the following entry in the `.mailrc` file:

```
set PAGER=metamail crt=1
```

to pipe each message (except the one-liners) to *`metamail`*.

Mail Forwarding, Sorting, and Automatic Reply

On occasion, you may want to arrange for all mail arriving to an account to be forwarded to another account or machine. For example, if you go on a sabbatical, you may prefer to receive the mail at your local account, rather than to log in via the network to the machine in your home organization. This can be accomplished by editing a `.forward` file in your `HOME` directory before you leave. In its simplest form, the `.forward` file contains the address to which you want all mail forwarded. You may also want to forward all mail from the accounts that you do not use daily to the account that you log into every day.

The `.forward` file has many more uses. Some of them may be restricted (for your own good) by the system administrator, but in principle, you can request that all incoming messages be automatically processed by a UNIX command or a program. The most popular of these is the *`vacation`* program, which is frequently found in the `/usr/bin` or `/usr/ucb` directory. This program is meant to return a short message automatically to people who send you e-mail when you are out of town telling them that their mail will be read after your return. Before you can use it, you need to create a message to be sent out as an automatic reply. Edit the file `.vacation.msg` and type the complete message (with fundamental header lines) that you want sent automatically. For example:

```
From: jkl@osc.edu (Jan Labanowski via vacation program)
Subject: I am on vacation until end of month.
Precedence: bulk

I am on vacation until Dec.31. If you have urgent matters,
contact my secretary Alicia Smith <alicia@osc.edu>.
Jan Labanowski, jkl@osc.edu
```

Once you create the `.vacation.msg` file, you need to initialize the program by running it manually with the `-I` (or for some versions with the `-i`) option. This step initializes (or clears) databases for the addresses of senders. The vacation program sends the automatic response from `.vacation.msg` file only once to a given address and saves this address in the database. Even if someone keeps sending you messages when you are out, the automatic response will be sent only once (as a precaution against loops and for protection of electronic lists). To install this answering machine place the following line:

```
\jkl, "/usr/ucb/vacation jkl"
```

in your `.forward` file. Of course, the `jkl` should be replaced with your own login name. This line will cause one copy of the message to go to your system mailbox, and another copy to be piped through the vacation program. This program extracts the `From:` header field from the incoming message and places the address in the `To:` field of the `.vacation.msg` message header. Then the automatic message containing the `.vacation.msg` is sent to the sender. The character \ in front of your name requests that the message go directly to your system mailbox and not be subjected to any other processing.

The `.forward` file can also be used to sort your mail automatically to different folders or even forward some messages automatically to other people depending on their contents or source. The *procmail*[2] program is one of the most popular tools for performing these chores. You can also have a popular *filter* program on your machine. The incoming mail can be sorted according to `Subject:`, the sender of the message, presence of some keywords in the message body, etc. The *procmail* program is very rich and flexible and offers an impressive spectrum of tools for automatic processing of your incoming mail. If you belong to several mailing lists and receive a high volume of mail, *procmail* or *filter* will save you a lot of time. The commands for *procmail* are entered in the `.procmailrc` file. As a result, all (or the selected portion) of your mail is automatically saved in designated folders. You can read these folders periodically (e.g., once a day) by invoking mail with an `-f folder-name` option, for example, the *mail* `-f folders/amber-mail` command.

Sending and Receiving Binary Files via e-mail

With the introduction of MIME, Internet e-mail is no longer restricted to sending text in English. Many MUAs can interpret the contents of the message according to the `Contents-Type:` header field automatically. A good overview of MIME is given in RFC-1521[3]. The MIME standard encourages extensions, and many new MIME types and subtypes have been adopted since its introduction. MIME types now exist for popular image, video, and sound files, as well as several word processor document formats.

An important portion of the MIME standard deals with encoding of the data. Electronic mail is guaranteed to pass only 7 bits of data. Without encoding, e-mail transport would most likely alter program binaries, binary data, images, etc. MIME provides robust ways of protecting data against alteration. Moreover, MIME enabled MUAs to perform encoding and decoding automatically and transparently, without need of user intervention.

However, old standards die hard, and sending binary files or foreign language text over the network can be done in many different and incompatible ways. Moreover, the ASCII standard for American English is not the only way to represent characters. Apart from national variations, there are alternative ways of encoding even English texts (e.g., the EBCDIC). The problem is aggravated by the fact that a one-to-one correspondence does not exist between some of these character encodings. When converting text from encoding A to encoding B and then back to encoding A, there is no guarantee that the original text will be restored exactly. The major difficulty is with symbols and punctuation characters that may not have counterparts in another encoding. Additional difficulty stems from the fact that some e-mail transport agents automatically convert tabs to spaces, remove trailing spaces, break long lines into short ones, and convert new-line characters. MIME addresses all of these issues and provides for fail-proof methods of encoding messages containing nontext or foreign language characters. However, many solutions that were invented before MIME are currently still popular.

The most popular pre-MIME mechanism for transporting binary files (or files containing text consisting of 8-bit wide characters, e.g., foreign language text or the IBM-PC Extended Character Set) are the *uuencode* and *uudecode* programs. If you get a message whose body is similar to this (lines start with `M`):

```
begin 644 test.exe
M5&AI<R!W87,@<W5P<&]S960@=&\@8F4@82!S96-R970@86YD(‘EO=2!W97)E
...many similar uuencoded-lines deleted...
M(&1I;&EG96YT(&5N;W5G: IT;R!T>7!E(&ET(&EN(&%N9”!S964@=VAA="!I
end
```

it is probably some binary file that has been uuencoded. To restore the original file, use the *uudecode* program, which is standard for UNIX, but which exists for many other platforms ranging from IBM PC to VAX. The uuencoded message may come in several parts, and you must splice the uuencoded pieces in the correct order by deleting headers and signatures. The *uudecode* program looks for the opening line, which starts with `begin`, followed by UNIX file permissions, and the name of the original file. Assuming that your e-mail message was saved in the file called `test.uu`, you may restore the original file (in this example `test.exe`) by executing `uudecode test.uu`. A good practice is to check if the file specified on the begin line does not exist on your system because the *uudecode* program would replace it with the new one.

If you need to send a binary file via e-mail, and you (or your addressee) does not have MIME capability, you may uuencode the file by issuing the command:

```
uuencode file1 file2 > file3
```

Here `file1` is the original binary file to be uuencoded, and `file2` is the name to be put on the `begin` line (it may be different from the name of the original binary file if you wish). The uuencoded version of the file is sent to the screen (standard output). To capture it, you need to redirect output to a file, in this case `file3`.

Many other encoders besides *uuencode* are available; however, *uuencode* is the most popular. For example, program *xxencode* is similar to *uuencode* (lines start with h):

```
begin 755 big.zip
hUEA-0k+-+++++2++++GQs++++++++0+U++++++++++0w20++o+CUE761c2GJ
...many similar xxencoded lines deleted...
hQaJkMKZmNKFiMKpZ+3xmNL-VOL7cNK3YNL6+LqxpR4FdMKRiPrBdQqldQrE+
hLr7ZQ43dQatVPKI+++++++++++++++++++++++++++++++++++++++++++
+
end
```

The corresponding decoding program is called *xxdecode*. The *xxencode* program was introduced because the *uudecode* program is not fail-proof, and sometimes, when a uuencoded message has to travel via machines using EBCDIC encoding (usually Bitnet nodes), it may be modified in such a way that the binary file recovered will not be identical with the original. The *xxencode* program was designed to fix this problem.

Another encoding program that you may encounter is called *btoa* (binary to ASCII), and the corresponding decoding program is called *atob* (actually they are usually the same program with two different names and, on the basis of the name with which it is called, it figures out what to do). An example of a *btoa* file is

```
xbtoa5 78 prog.zip Begin
JHGo\!!*'"!!#7a!!JhMz!!",azz]FZ1cci]q+Ooq@FPq!mXPQIK\PQF)W(B=G+e0<oQ!rr<,_>n$j
NJ<0a3!e:7R!<<*"5QCck!<<*"5QF%P!<<*"5QCl'S0O[L5QF%O!<<*"5QF%P!<<*"S`@Iu0)ttXXJ
...many similar lines of btoa-encoded message deleted...
xbtoa End N 81920 14000 E 1 S 68d1bf R 55dc4039
```

To encode a binary file use `btoa binfile encodedfile`, and to decode it, use `atob encodedfile`. For the Macintosh platform, the *BinHex* program is the most popular. Files encoded with this program are usually marked with the extension `.hqx`.

An encoded file is always larger than the original binary file (unless it was also compressed). Encoders usually work in such a way that they combine several binary characters into an aggregate and then chop this aggregate into smaller pieces, each piece having only a limited range that can be represented by popular letters and punctuation marks. For example, the *uuencode* program combines three 8-bit characters (each capable of representing up to 256 different character codes) to produce a 24-bit aggregate that is then sliced into four 6-bit pieces (which can each represent only 64 different codes). These 6-bit pieces are converted into corresponding popular alphanumeric and punctuation characters and sent via e-mail. On the other end, the *uudecode* program converts these characters back to the 6-bit pieces, four of which are combined into a 24-bit piece that is then chopped into 8-bit codes corresponding to the original bytes of the binary file.

Another way of encoding is called quoting. In this case, the characters that are not ASCII are represented as their codes (usually hexadecimal codes). The quoting character is usually an = sign (which also needs to be quoted). For example, you could find the following fragment of text: "I graduated from the Universit=82 de Montr=82al a year ago and moved to M=81nchen."

You may also find that the files were compressed before encoding. Many programs can substantially reduce the size of the file by converting it to a binary file. The techniques for file compression are beyond the scope of this chapter; they simply pack the information more efficiently. Written English text does not use more than about 50–60 characters, and each character still occupies one whole byte (i.e., a unit that can represent 256 different characters). Possibly this space can be recovered, for example, by packing four letters into 3 bytes and saving 25% on volume. Moreover, some patterns are more frequent than others and can be encoded economically. The resulting compressed file is always binary, even if the original was not, and for transmitting over e-mail, the file needs to be encoded. Compression may shrink a file up to 5 times and is frequently used to save disk space and bandwidth.

Many compressing and decompressing programs are available, for example: the UNIX *compress/uncompress* (the `.Z` extension is used to mark compressed files), *pkzip/pkunzip* (the `.zip` extension is used), *arj/unarj* (extension `.arj`), and *gzip/gunzip* (extension `.gz`). These programs are popular in many archives and are available free to anybody connected to the Internet either as a source code or versions compiled for popular computers.

References

[1]A utility for processing multimedia mail written by Nathaniel S. Borenstein of Bellcore, available from anonymous FTP **thumper.bellcore.com** in `pub/nsb`.

[2]Program for sorting mail written by Stephen R. van den Berg at RWTH-Aachen, available from **ftp.informatik.rwth-aachen.de** as `/pub/packages/procmail/ procmail.tar.Z`.

[3]Borenstein, N.; Freed, N. "MIME (Multipurpose Internet Mail Extensions) Part One: Mechanisms for Specifying and Describing the Format of Internet Message Bodies", 1993, RFC-1521.

chapter four

Electronic Lists

Jan K. Labanowski, Charles F. Bender, and Alejandro Pisanty

In This Chapter

This chapter discusses the different kinds of electronic lists and gives instructions for subscribing, unsubscribing, and posting to them.

3223–7/96/0065$12.00/1

SOMETIMES YOU NEED TO SEND INFORMATION to a group of people rather than an individual. Electronic mail (e-mail) provides an alternative to the copying machine and mass mailings. It also has the important advantage that information is distributed quickly. The two possible scenarios of mass distribution of e-mail are unidirectional and bidirectional. The distribution list, also called a mailing list or electronic news service, is an example of unidirectional electronic distribution, that is, only a single person is sending material to the recipients. An example of a bidirectional list is the electronic discussion forum, in which recipients may also distribute their own messages to other recipients. The electronic discussion forum is frequently called a bulletin board, a mail reflector, a mail exploder, an electronic list, or a list server (often abbreviated to a listserv jargon term). There are no precise distinctions between these terms.

Electronic mail is not the only medium for sending information to groups of people. Unidirectional electronic information can also be distributed via

- the World Wide Web (WWW), which allows presentation of multimedia HyperText documents (including text, graphics, audio, and video) to computers connected to the Internet.[1] Documents are written in HTML (HyperText markup language) and can be linked together. To access this information, the computer needs to have Web browser software installed (e.g., `Mosaic`, `Netscape`, or `Lynx`). Chapter 6 discusses the World Wide Web in detail.
- Gopher, a predecessor of the WWW that is essentially a document-viewing protocol, although it provides also for images and database-searching capabilities. Gopher documents cannot be linked together as in the Web. Gopher sites are now usually accessed with Web browsers, as they uniformly provide support for this protocol. Chapter 5 discusses Gopher in detail.
- the read-only Usenet newsgroups that are most frequently organized as a news service (e.g., ClariNet).

Bidirectional electronic information can be distributed not only via e-mail but also via

- Usenet, a protocol for maintaining discussion groups on the Internet. It is similar to e-mail, but the messages are not targeted at a particular address or user but can be accessed by all users who have accounts on the connected computer. The discussions are organized hierarchically as newsgroups, and special software, called a newsreader (e.g., `rn`, `tass`, and `rrn`) is needed to read and navigate among thousands of available newsgroups.

- FTP, an old but still the most popular protocol for exchanging larger bits of information, usually large documents, data files, or software libraries and programs. Its uploading option is frequently disabled to prevent the storage of unwelcome information (e.g., explicit images or other controversial materials), but in principle it was envisioned as a bidirectional exchange medium.

For bidirectional dissemination, Usenet is the most popular medium. In many cases, however, e-mail is advantageous for distribution because the list of recipients is defined and closed.

Small Personal Lists

The distributions can be organized in many ways. If you need to forward mail to a few people (e.g., your research group), in UNIX it is easiest to set up an alias in your `.mailrc` file. This solution is satisfactory if the messages are unidirectional. However, if you want to organize an electronic discussion forum in such a way that messages can be bidirectional, each member of the discussion circle must have the same alias in his or her mailbox, and it must be current.

A much better solution is to use the tools provided by `sendmail`. An integral part of `sendmail` is a database of aliases. Its readable form is present in the file `/etc/aliases` on UNIX machines (although the location and name of this file can be changed at installation). This file contains entries in the form `some-alias: addresses`, and the entries can have many different forms. The simplest form is just a single e-mail address (or an alias) on the local or remote machine. However, an entry may also be a list of addresses separated by commas, for example,

> orgchem: paul, smith@far.uosx.edu, barb@lclc.org, richard, muerl, ada@phenol.com, stankuch@ufla.bub.net, gadocha@retired.ball.team.com, gradstudents

Each message sent to the **orgchem@the.machine** address will be automatically forwarded to all the addresses on the list. Of course, the addresses need not point only to computer accounts but can also be aliases themselves (e.g., gradstudents in the orgchem example). You can organize a hierarchical list this way; that is, you can nest aliases within aliases. You need to be careful, however, not to create loops (e.g., if gradstudents contained orgchem as one of its addresses). You may also point to other lists. It is especially useful if the lists are located on remote machines. In this way only a single message needs to be sent to the remote site, and remote lists can be administered by their local administrators.

The `/etc/aliases` file is a very sensitive item because of computer security, so it can be modified only by privileged users. Placing larger and frequently changing lists as an explicit alias in the `/etc/aliases` file is not practical. A better way is to use the `include` facility of *`sendmail`*, which allows the addresses to be placed within a file. This file should be made accessible to the person who will keep the list current. For example:

```
orgchem: :include:/home/paul/orgchem-members
```

and the corresponding file `/home/paul/orgchem-members` may look like

```
"Paul McMarty - ListAdmin" paul,
"John Smith - Univ of S. Xanandoga" smith@far.uosx.edu,
"Barbara Novak - LCLC Member" barb@lclc.org,
 .........
"J Gadocha - Old Friend" gadocha@retired.ball.team.com
```

Such a list can be changed at will by somebody who is an administrator of the list rather than a system administrator of the machine.

A standard practice is to provide additional addresses that should be used for messages intended for the administrator only, such as subscriptions, unsubscriptions, or address changes. For that reason, the complete entry in the `/etc/aliases` file should look like

```
orgchem: :include:/home/paul/orgchem-members
owner-orgchem: paul
orgchem-request: paul
```

The **owner-SOMELIST** address is used automatically by many systems to report mail delivery errors (e.g., when the account has been canceled or machine decommissioned). Such messages obviously should go to a list administrator rather than the whole list. Unfortunately, it may happen that the mail bounces will be directed to the original sender of the message rather than an **owner-SOMELIST** address, because some machines do not adhere to this convention (the whole story is more complicated, but we will leave it at that). The **SOMELIST-request** address is intended for administrative messages requesting subscription, unsubscription, or address changes. These messages will not go to the whole list, but will be relayed to the person who is in charge of administering the list. This simple list organization is suitable for small lists.

Large Electronic Lists

For large lists, with hundreds of subscribers, this scheme is not good enough. Before you undertake a task of starting a large list you should understand the issues involved in managing this resource (a good introduction is available[2]). Imagine that some malconfigured machine (and there are plenty of them) starts sending "bounced" messages (those that are not able to be delivered for one reason or another) to the list rather than its administrator. It will respond to each message it receives with a new bounce, that is, by creating a mailing loop. In the era of fast networks, it will produce several hundreds messages within a few hours. If the list administrator is not currently at his or her terminal, the results may be devastating; hundreds of pieces of junk electronic mail may go in the mailboxes of each list subscriber.

For that reason, larger electronic lists are serviced by special software that performs many checks before a message is released to subscribers. The only fail-proof method to prevent mailing loops from happening is to have a person review each message before forwarding it to subscribers (called a moderated list), but good list-management software will in most cases prevent mailing loops. Many lists are administered with software that is developed in-house. However, several excellent publicly available packages assist in list administration. The names and descriptions of popular list-administration software are given in the Appendix.

The lack of standardization in this field is a mixed blessing, because although all of them use similar commands, some differences exist. The best way to find out about the valid commands is to send a single line of text containing `HELP` to the main address of the list server. It is usually **LISTSERV@ some.machine** but it may be **majordomo@some.machine**, **almanac@some.machine**, **listproc @some. machine**, **mailbase@some.machine**, or even **mailserv@ some.machine**. A global address may not be associated with all the lists on the **some.machine**, and in this case you need to contact the administrator of the given list at the address **listname-request@**

some.machine. If this does not work, try **owner-listname@ some. machine**.

For Internet-style lists, the **listname-request@some.machine** address is a standard way to contact the list administrator, but Bitnet lists usually do not have such aliases defined. Typical Bitnet list names are restricted to eight characters. Never send requests of administrative type directly to the list because you are not likely to get any polite or informative responses. If you are using automatic services of the list-administration software, remember to put your commands in the body of your e-mail message. The `Subject:` field is not usually scanned for commands.

How To Subscribe to Any List

Generally, to subscribe to any list you need to send the following message:

```
SUBSCRIBE listname Firstname Lastname
```

to the address of the list-managing software, for example,

almanac@some.machine
ListProc@some.machine
LISTSERV@some.machine
majordomo@some.machine

As usual, there are exceptions to this rule. For example, one of the largest sites for electronic lists, **MAILBASE@ mailbase.ac.uk**, suggests that you subscribe by sending the message

```
JOIN listname Firstname Lastname
```

to MAILBASE@mailbase.ac.uk.

If these methods do not work, skip the name portion; that is, send the message

```
SUBSCRIBE listname
```

and try again. If it still does not work try to contact list the administrator at **listname-request@some.machine** or the person in charge of mail, **postmaster@some.machine**.

On joining the list you usually get detailed information about list services and rules. It is a good idea to save this information for future reference.

How To Unsubscribe from Any List

Send the message

```
UNSUBSCRIBE listname
```

to the administrative address of the list-managing software. For LISTSERV lists, use the

```
SIGNOFF listname
```

command. For MAILBASE lists, use

```
LEAVE listname
```

How To Post to a List

Many lists allow only subscribers to post to the list. Posting is done by sending your message to the distribution address: **listname@some.machine**

General compilations of electronic lists are available. Diane K. Kovacs (address **dkovacs@kentvm.kent.edu** or **dkovacs@kentvm.bitnet**) periodically releases the Directory of Scholarly Electronic Conferences, which can be retrieved by sending a message with the lines:

```
GET ACADLIST FILEn (n = 1 to 9)
GET ACADLIST INDEX
GET ACADLIST.README
```

to **listserv@kentvm.bitnet** or **listserv@kentvm.kent.edu**. Another general index of Listserv electronic lists can be retrieved by sending the message `GET LISTSERV LISTS` to **listserv@bitnic.bitnet**. Yet another list can be retrieved from **rtfm.mit.edu** either via anonymous FTP or via e-mail from the directory `/pub/usenet/news.answers/mail/mailing-lists`.

References

[1]Krol, E. *The Whole Internet;* O'Reilly & Associates: Sebastopol, CA, 1994.
[2]Peek, J. *Managing Internet Information Services;* O'Reilly & Associates: Sebastopol, CA, 1994.

Appendix: Popular List-Administration Software

`Almanac` written at Oregon State University is a list-management system written in C. It is highly configurable and supports automatic subscription, unsubscription, and file retrieval via e-mail. It can be obtained by anonymous FTP from **oes.orst.edu** as: `/pub/almanac/almanac-1.5.1b. tar.Z`. To subscribe to a list handled by `Almanac` send the message `subscribe listname` to the address **almanac@some.machine**. The address almanac is reserved for administrative purposes, and the listname is a name of the list. You can get a catalog of available lists by sending the message `send mail-catalog` to **almanac@ some.machine**. To unsubscribe, send the message `unsubscribe listname`. Your e-mail address is taken from the header of your message. To post to the list, just send your message to the address **listname@some. machine**.

`Listserv` is a standard software package for IBM mainframes. It is probably the oldest and the best-known tool for list administration, and for that reason many newer tools imitate the syntax and features of this system. `Listserv` supports file retrieval via e-mail, automatic message archiving, subscription and unsubscription, information services, and archive searches. Some `Listserv` clones for UNIX do not support all these features. All administrative tasks and commands to the server are sent to the address **LISTSERV@some.machine**. To get general help from a list server, send a single line, `INFO` or `HELP`, in the body of the message. To get a list of electronic lists distributed by the given list server, send a single-line message consisting of the word `LISTS`. To obtain detailed information about a given list, send the request `REVIEW listname`, where `listname` is the name of the list you want to review. To retrieve the list of files associated with a given list, send the message `INDEX listname` or `GET list-`

name FILELIST. Once you have the list of available files, you can retrieve them by sending messages like GET filename filetype, where filename is usually the same as listname and filetype is the name of a specific file. You can even search the files in *Listserv* archives for keywords. To subscribe to an electronic list, send the message SUBSCRIBE listname Firstname Lastname, where listname is the name of the list that you want to belong to. You are also required to enter your name. To unsubscribe, send the message SIGNOFF listname to the *Listserv* address. If you are unsuccessful with automatic unsubscription, you may want to review the list of subscribers by sending REVIEW listname. You may find that you were subscribed from a different machine and the mail is forwarded to your current account. To unsubscribe from the list, you need to send the message from the same machine from which you subscribed. Many clones of the *Listserv* program for non-IBM machines exist, and the best-known is the *ListProcessor* software described as follows.

ListProcessor, also called *ListProc*, written by Anastasios Kotsikonas, is a very extensive and flexible list-management program that also supports hierarchical distribution, list moderation, archive searching, file retrieval via e-mail, and more. It is available by anonymous FTP: **cs-ftp.bu.edu** in the directory /pub/listserv. *ListProcessor* supports all the basic commands of the original *Listserv*; that is, you subscribe by sending a message SUB listname Firstname Lastname to **LISTPROC@some. machine**. This address is sometimes also aliased as **LISTSERV@some.machine**. To unsubscribe, send the message UNSUB listname or SIGNOFF listname. You can also check to which lists you are subscribed by sending the single word WHICH as a message to **LISTPROC@some.machine**. This list-management software is well-suited for managing big lists; however, it may be overkill for a small list.

Majordomo is a simple list-management software written in Perl language by Brent Chapman. It supports subscription and unsubscription without list administrator intervention. To subscribe and unsubscribe, the command syntax is subscribe listname and unsubscribe listname, respectively. These commands must be sent to the address **majordomo@some.machine**. You can also obtain files by using the get listname filename command, get an index of available files with the `index listname` command, check which lists are available by sending a single word, `lists`, retrieve subscribers of the given list by sending `who listname`, and check which lists you are subscribed to by sending the single word, `which` to the address **majordomo@some.machine**. The latest version also provides tools for list moderation and message approval. It is simple and well-suited for small lists. The latest version can be retrieved by anonymous FTP from **ftp.greatcircle.com** in the directory `/pub/majordomo`.

`SmartList` is a relatively new addition to list-management software written by Stephen R. van den Berg (who is also the author of the excellent `procmail` package). This software is available by anonymous FTP: **ftp.informatik. rwth-aachen.de** as: `/pub/packages/procmail/ SmartList.tar.Z`. It provides a lot of modern features and has a built-in intelligence. For example, it does not require a rigid format for subscription messages, and it will redirect the subscription messages away from the electronic list. It provides MIME support for list archives and allows for remote list moderation. It is not popular yet, but sets the standard for the future generation of list-management software.

chapter five

Gopher

Steven M. Bachrach

In This Chapter

This chapter shows how the network "communication problem" led to the development of Gopher. It also discusses how to use the Gopher client and how to set up and operate a Gopher server. Finally, the future of Gopher and how the chemistry community can make further use of this Internet tool are discussed.

3223–7/96/0075$15.00/1

THE INTERNET IS, at its most basic level, a collection of computers with their associated files connected by wires. The valuable asset in this assembly is the files, particularly the information contained in these files. Organizing, sorting, and indexing these files is the key step for making them useful to users around the world. A secondary concern is how to enable users to obtain the files they need—how does a user transfer the file from, for example, an IBM UNIX workstation halfway around the world, to a Macintosh computer? Organization and transmission of data is what we will call "the communication problem".

The Communication Problem

As the Internet has evolved, it has become very heterogeneous in many ways. The many different types of computers range from microcomputers to workstations to mainframes to supercomputers. The many operating systems include DOS and Windows, dominant within the PC community, and UNIX, dominant in the workstation genre. The connection between computers can also be of many different types. Finally, data files are written by an increasing number of software programs, many of which have their own format. Considering all the possible combinations of these factors, it is amazing that the Internet even exists, let alone functions!

Fortunately, a number of standards, called ***protocols***, have been established to facilitate the communication between computers. The most important of these protocols is TCP/IP (transmission control protocol/Internet protocol), which dictates how data is moved between computers. We will forgo most of the details of TCP/IP, except for two aspects: addressing and FTP (file-transfer protocol).

Addressing is the means for giving every computer its own unique name. As an analogy, the postal service would have extreme difficulty in delivering mail to you if your home did not have a unique designation. The same problem exists on computer networks. The Internet protocol defines a procedure for naming every computer. Addresses are assigned as a combination of four fields (called ***domains***) that are each 8 bits long (allowing for numbers ranging from 0 to 255). These four fields are separated by periods; the generic address is **aaa.bbb.ccc.ddd**. An example is the Northern Illinois University (NIU) Chemistry gopher site; its address is **131.156.9.16**. Each domain defines an ever more specific locale. The set formed by the first two domains (aaa.bbb) defines the organization; in this case 131.156 uniquely designates Northern Illinois University. The next two domains define departments and individual computers in the department; the 9 indicates the chemistry department, and 16 indicates the specific computer in the chemistry department.

Using numbers is the preferred method for identifying computers to *other computers*, but people tend to communicate better with words. The Internet protocol allows for a more descriptive analogue of addressing, using words instead of numbers for the fields. The NIU Chemistry gopher site has the address **hackberry.chem.niu.edu**. The main difference is the order of domains. Here the last domain is the least specific; edu indicates an educational institution. Next to last is the institution name (niu in this case), and the department name is in front of that (chem). The first domain is the local name for the computer, hackberry in our example. These two addressing systems are completely interchangeable and synonymous. You are free to use the number or name convention at your own discretion. However, to communicate with another computer, you will need the address of that host computer. If you frequently connect to many different computers, keeping track of all the addresses can be a challenge.

The principal means for file exchange on the Internet is through a utility called FTP, which actually designates both a software program and a protocol. The FTP utility allows a user to establish a connection between his or her computer (client) and a remote computer (host) and then transfer files between the two. This connection requires that the user know the address of the host and have an account on the host. A system administrator can avoid the problem of individuals needing an account by setting up an anonymous FTP account. The anonymous FTP account allows the user to log in as "anonymous" and then give his or her electronic mail (e-mail) address as the password. Once connected, the user must know a number of fairly arcane UNIX-based commands to navigate file systems and send or receive files.

Now, none of this procedure is particularly difficult, but a learning curve is associated with any computational task. A procedure that would allow users to connect with computers without knowing addresses and allow file transfers with simple commands (like a mouse-click or a menu selection) would certainly streamline the process.

Once connected to a remote computer, the truly difficult task begins—how to locate the files of interest. FTP allows you to move through directories and examine file names. If the system administrator is exceptionally diligent and dedicated, the directory structure may provide some level of useful organization. If file names are descriptive, users may be able to discern enough information to locate the files they require. Some computer systems place drastic limitations on file-name length, so names may not be of significant help. Most file systems are not well organized.

Furthermore, this directory-based structure allows for no search capabilities whatsoever. If you are attempting to locate all files on a certain topic, and they do not happen to all reside in a particular directory, you probably will not be able to locate them all.

Although FTP is a powerful tool, particularly in its ability to facilitate file transfers across multiple platforms, it is too rigid and arcane to be useful to nonexpert computer users. The lack of file structure and search capabilities limits the use of FTP as an information delivery system. The Gopher client–server software was developed precisely to address this communication problem.

The Gopher Solution

The communication problem was the driving force behind the development of *Gopher*, which was written at the University of Minnesota. (Thus the origin of the name—the mascot of the University of Minnesota is the Golden Gopher. The name is actually a pun on the word gopher, as the function of the program is to *retrieve* files.) It is a client–server navigation system. The server software resides at the remote (host) computer and transmits the requested files to the client. The client software resides on the local computer. It handles the operations for connecting to the gopher servers, sending requests to the server, and displaying files for the user.

The Gopher server requires files to be organized in a tight directory structure. Using an adjunct program, the user can index the files and make them available for searching within the Gopher client–server framework. The Gopher server can deliver a variety of document types, including connections to other Gopher servers. With well-designed servers, a user need know only the address of the initial server; the addresses of all other related servers are provided. The client software presents a menu-based user interface for file selection. No knowledge of any transmission commands is required. Therefore, the communication problem has been eliminated.

The Gopher Client

Most users will experience Gopher from the client side only. The client is software that resides on your local computer and acts as your gateway to GopherSpace, the network of all Gopher servers. The client establishes the connection to the server, prepares the menu-based interface, sends all requests to the server, receives the transmitted files, and displays these files on the screen. Even though you may in a typical session connect to many different types of computers around the world, the user interface will remain uniform. You need to supply an address on the initial Gopher connection, and often you will not even know what server you are connected to at any particular time.

Setting Up the Gopher Client

The Gopher client is a software tool that can be obtained free of charge. You can obtain the client via anonymous FTP to **boombox. micro.umn.edu** and look in the `/pub/gopher` directory for the subdirectory that best fits the description of the computer you will be working on. There are versions of the client available for UNIX workstations, PCs, and Macintosh computers. Other anonymous FTP sites that have the Gopher software are listed in Table 5.1.

UNIX Client. The UNIX client will need to be compiled after you obtain it. The downloaded code actually contains both the client and the server. Examine the `README` file to obtain the latest information on compiling the program. Make sure to check the file `makefile.config` for any specific changes for your computer. In particular, make sure the PREFIX variable is set to the path where your source files are located.

Table 5.1. Locations of Gopher Software

Address	*Directory*
boombox.micro.umn.edu	`/pub/gopher`
src.doc.ic.ac.uk	`/public/computing/archiving/gopher`
ftp.uni-kl.de	`/pub0/gopher`
nic.swithc.ch	`/mirror/gopher`
bash.cc.kcio.ac.jp	`pub/mirror/gopher`

d to edit the file conf.h. Specifically, you need to ıddresses of two gopher servers to connect to when nt. (These can be overridden, if desired.) The vari- 'OST and CLIENT1_PORT define the address and port number of the first server to try, and CLIENT2_HOST and CLIENT2_PORT define the second server to try, if the first is unable to respond. The HOST variable is the address of the Gopher server you wish to connect to by default. If you do not know of any specific gopher that you want to use as a default, we list in Chapter 11 a number of chemistry-related gopher sites and their addresses, and any of these can be used as a default host. We will discuss what the PORT variable means in the section describing the server, but most Gopher servers use port 70.

Once the Gopher client has been successfully compiled, it is invoked with the command `gopher`. Typing this command will bring up the client and automatically connect you to one of the default gopher servers. If you wish to directly connect to a different server, use the command `gopher address port`, where `address` is the address of the server and `port` is its port number.

PC Client. The PC client is available for DOS 3.3 or higher and requires 640 kilobytes (K) of memory. (Windows versions are also available.) It is available in either a *zip* version (which is compressed code that must be uncompressed by using the *pkzip* program) or as a self-extracting archive. The self-extracting archive is the easier to use; simply type the file name (`pcg3binx.exe`), and it will automatically uncompress and place the gopher files in your directory. Before you execute the client, make sure to read the `pcg3.txt` file for the latest instructions. In particular, note the information on setting up your network connection properly.

The client has a configuration file that holds user-specific data. To modify this file, run the client by typing the command `gopher` and then open the `Application...` submenu under the `Configure` menu. You should enter the address of the default server you wish to connect to in the Home Gopher Server field, with its port number in the Port Number field, and you can add a second server to the Alternative Gopher Server field. Make sure to supply the appropriate information for your particular hardware, software, and network configuration in the other fields.

Once the configuration has been completed, you can start the client by typing the `gopher` command. You will automatically be connected to one of the default servers you listed in the configure file. If you wish to connect to another server, use the New Gopher submenu in the File menu.

Macintosh Client. The Macintosh client is called TurboGopher. MacTCP must be running on your computer. Make sure that MacTCP has been set up properly, especially the name server information. TurboGopher is available as compressed file that can be decompressed using the program *`Stuffit`*.

TurboGopher is invoked by double-clicking on its icon. The program will default to connect to the University of Minnesota Gopher server. This default can be changed by selecting the Configure TurboGopher submenu in the Setup menu. You can enter the address and port number for two servers. Once the default is established, every time you subsequently start TurboGopher, you will connect to one of the default servers. You can always connect to another server by selecting Another Gopher in the File menu. Enter the address and port number and click on OK.

Other Clients. Gopher servers can be accessed using nongopher clients that are aware of the Gopher system. All World Wide Web clients (frequently called ***browsers***) can be used as Gopher clients. These browsers are discussed in Chapter 6.

Additionally, some of the commercial on-line services, such as Genie and America Online, offer Gopher access. These services provide clients that are very similar to the ones just described.

Using the Gopher Client

Exploring GopherSpace. Once you have connected to the default Gopher server, the client will present you with a menu of options supplied by that server. Selections from the menu are made by placing the cursor next to the item desired using the arrow keys and pressing the Return or Enter key (for UNIX and PC clients) or by double-clicking on the item (for PC clients with mouse and Macintosh clients). The menu delivered by the Northern Illinois University Chemistry Gopher, as viewed using TurboGopher, is shown in Figure 5.1. Figure

Figure 5.1. NIU Gopher server home menu viewed using TurboGopher on a Macintosh.

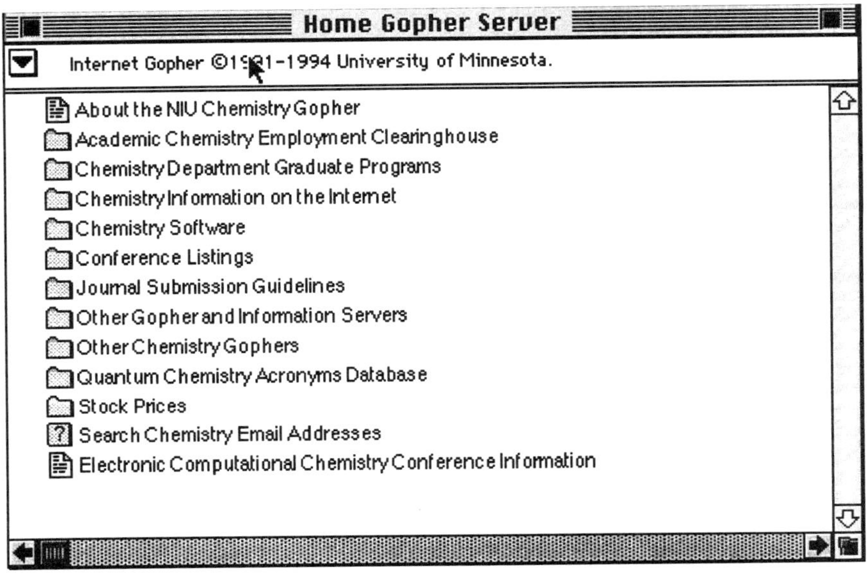

5.2 shows the menu from the `Chemistry Information on the Internet` directory in the NIU Chemistry Gopher, viewed using the UNIX client. Gopher can deliver a variety of different types of information, which we will now discuss.

Text files make up the bulk of the information presented by Gopher. Text files are indicated in the menu by just the file name when using the UNIX client, with a preceding `<F>` on the PC client, and with an icon of a printed page on the Macintosh client. When you select a text file, that file is displayed on your computer screen. TurboGopher will actually open a new window with the document displayed within it. When you are finished with the document, you can return to the previous screen (the menu) by typing the letter u with the UNIX client or clicking on the appropriate window with the Macintosh client.

The menu may contain folders or subdirectories, which can themselves contain a variety of different types of information. A directory is indicated by a slash (/) at the end of the file name by the UNIX client, by a preceding `<D>` on the PC client, and by an icon of a folder on the

Figure 5.2. The `Chemistry Information on the Internet` menu of the NIU Chemistry Gopher site viewed using a UNIX client.

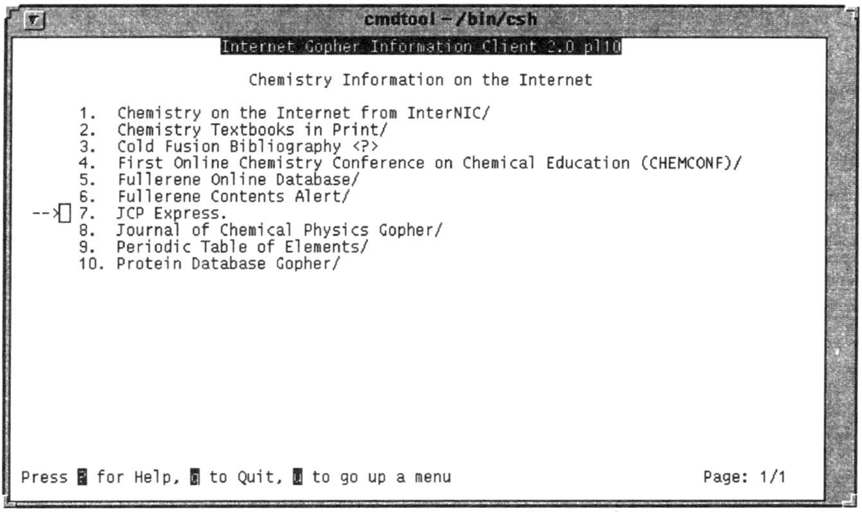

Macintosh client. When you select a directory, the server supplies a new menu screen containing the items for selection held within that subdirectory.

The Gopher server may have the ability to search files. The search item is indicated by <?> at the end of the name on the UNIX client, by a preceding `<S>` on the PC client, and by an icon of a page with a question mark on the Macintosh client. Searches are typically handled by a program called WAIS (Wide Area Information Service). WAIS searches are discussed in a later section.

Gopher can initiate remote sessions with other computers using the `Telnet` facility, if it is available on the local computer. The Telnet item is indicated by `<TEL>` at the end of the name on the UNIX client, by a preceding `<T>` on the PC client, and by an icon of a computer on the Macintosh client. Selection of the Telnet item initiates a remote login session with a new host computer. Instructions for logging onto the remote computer are usually supplied so that anyone can take advantage of the feature.

Presenting documents maintained in an ordered directory structure is a very nice feature, but the key to the rapid growth of Gopher servers is the ability to link the servers together. Instead of duplicating material, a menu entry to another gopher server allows the user to seamlessly disconnect from the initial server and connect to the new server. The address of the new server is provided by the information contained in the link; the user never supplies any data on the new server. In fact, the user may not even be aware of what Gopher server is supplying the current menu. Links can be made to a directory or a file on the remote computer. Links are not given a special icon by the clients, rather they are treated as just another text file or directory that just happen to be located at a different site. The user simply selects an item of interest. The new connection is made transparently, and either a file or a new menu is displayed.

Because links are easy to add to a server, their use is widespread, and hence many interconnections exist between Gopher servers. These interconnections lead to the term GopherSpace, which is the collection of all Gopher servers that are connected by links. Users "tunnel" through GopherSpace by making menu selections that connect them to computers around the world.

Bookmarks. Navigating through Gopher sites to get to the server, directory, or file of interest is not difficult. It can, however, be arduous, particularly if your path takes you through many different servers. Once you have located a site of interest, Gopher does provide a means for creating a ***bookmark***, which saves the information needed to directly access that site. The next time you wish to visit that site, you simply select the bookmark and you will be directly connected to the desired site.

Bookmarks are created very simply by the client program.

- Using the UNIX client, navigate to the site for which you wish to create a bookmark. If this site is a server or directory, then type `A` to create the bookmark; if you want save the location of a file, then type `a`.
- On a PC, navigate to the site you wish to save. Then select `Open Bookmark File` from the `File` menu. The bookmark list will open up; click on the add button and enter the name you wish to give to this site.
- Using TurboGopher, once you have navigated to the site of interest, you can create bookmarks by selecting `Set Bookmark` in the `Gopher` menu. These entries can be edited by first highlighting the entry in the bookmark list and then selecting Edit Bookmark from the Gopher menu.

Using bookmarks is a very simple process.

- To access the bookmark list on the UNIX client, type `v`, which brings up a menu of all your bookmarks. Select one of these and you will directly connect to that site.
- To use the bookmarks on the PC, open the bookmark list using the `Open Bookmark File` in the `File` menu and choose the site you wish to visit.
- On the Macintosh, open the bookmark list by selecting `Show Bookmarks` from the `Gopher` menu and then double-clicking on the name of the site you desire.

Downloading Files. When a file is selected in the Gopher client, the file is transferred to the local computer and displayed on the screen. The file is not directed to the disk and as such is not available for reuse. The client can, however, save the files to disk and thus act in the traditional sense of downloading files.

The currently viewed document can be saved, using the UNIX client, by typing `s`. You will be prompted for a file name. With TurboGopher, the currently viewed text can be saved by selecting the submenu `Save as Text` from the `File` menu. To save the currently viewed file to disk using the PC client, select `Save File` from the `File` menu. Enter the file name and directory that you wish to save the file in. For very large files for which the PC has insufficient memory, the client will directly store the file to disk, after prompting for a file name and directory, and will not put the text on screen.

The Gopher Server

The key element to the Gopher system is the server residing on the host computer. The use of the system is dependent entirely on there being valuable information, resources, and tools available on the servers. In addition, the information must be held in a logical fashion so that users can locate the files they need. To this end, a simple search engine must be available to select only the desired files.

The burden of this function resides entirely with the administrator of the server, but the server software makes the job fairly easy. The system administrator does not need to be concerned with protocols and the details of interconnecting and transmitting files; these are built into the server. Rather, the administrator needs just to understand the file structure and methodology for establishing links to create a working gopher site. This section outlines the basics of operating a gopher site including the search tool WAIS, an alternative server program called GN, and the next generation of servers called Gopher+.

Setting Up the Gopher Server

This section tells you how to create a gopher server on a UNIX computer. Gopher servers can and are operating on PCs and Macintosh computers; however, the large size of documents (particularly graphics and movies) and the growing number of Gopher users place a premium on the need for speed and disk availability on a server. Establishing a server on non-UNIX platforms will require similar steps to those described here.

The Gopher software is available by anonymous FTP to **boombox.micro.umn.edu**. The files are kept in the `pub/gopher/Unix` directory. The files may be compressed and will certainly need to be restored using the `tar` command, such as `tar -xvf filename`.

Configuring the Server Software. Before compiling the server for your hardware, examine the `Makefile.config` and `config.h` files for any changes necessary for your particular computational environment. It should not be necessary to make any changes in any other files. Only two variables must be modified in all cases; both are in the `Makefile.config` file. The SERVERDATA variable defines the path for the `root` directory for your server. The `root` directory is the location of the menu items that will be served first when a connection is made to your Gopher. On the NIU Chemistry Gopher, the `root` directory is `/gopher/data`.

The SERVERPORT is the default TCP port for your server. The standard port for gopher servers is 70, but you may use any port that is convenient and not used by any other service. You must be a superuser to start a gopher server with `root` 70. These two variables can be overridden in the start command line, if you need to change them in the future and do not want to recompile the code.

Once the program has successfully compiled, the server program, called *gopherd*, will be placed in the `gopherd` subdirectory. The server can be initiated (from the superuser account) with the command `gopherd directory port`, where `directory` is the `root` directory and `port` is the port number. This command will initiate the gopher server, but the server will now be running with `root` access, and this can be a security problem. The preferred way to initiate the server is with the command `gopherd -c -u username directory port`. The `-c` command disables the chroot() default and the `-u` command runs the server out of the username account instead of the default `root` account.

Another feature that is quite helpful is to log the user access of the server. The resulting log file can provide statistics on the use of the server. To create the log file, include the command `-l logfilename` with the `gopherd` command when starting the server. The full path name of the log file is `logfilename`.

Once you have started the gopher server, you can test its operation by attempting to connect to it using a gopher client or by Telnet to the site (type `telnet gopher port`). The Gopher server should reply with the `.cache` file, which is described in the next section.

To ensure that the gopher server will restart whenever the computer is rebooted, you should add the start-up command line to the appropriate file in the `rc.local` or equivalent directory. For example, the `/etc/rc2.d/S93gopher` file at the NIU Chemistry Gopher is shown in Figure 5.3.

Adding Files to Your Server and Making Them Available. Once you have configured the server properly, the next step is to start adding data (files) to your server. A small digression into how the server makes files publicly available will help you to design a useful gopher server.

Figure 5.3. A file to start Gopher upon booting the computer.

```
case $1 in
'start')
  if [ -f /local/bin/gopherd ]; then
     /local/bin/gopherd -c -u admin -l /local/logs/gopher.log /gopher/data 70
  fi
'stop')
          ;;
'stop')
 pid=`/usr/bin/ps -e | /usr/bin/grep gopherd | /usr/bin/sed -e 's/^ *//' -e 's/ .*//Ŏ`
     if test "$pid"
     then
          kill $pid
     fi
     ;;
*)
     echo "usage: /etc/rc2.d/S93gopher {start|stop}"
     ;;
esac
```

Everything placed in the root directory, or any subdirectories under the root, is made available by the server. The program actually uses a file called .cache, which contains the information on all files and subdirectories in the current directory. The .cache files are generated automatically by the server when needed by creating one entry per file. Each entry contains the type of file, the name of the file as displayed in the menu, the path name of the file, the server name, the port number, and the MIME (multipurpose Internet mail extension) type, each delimited by a tab. An example .cache file (for the root directory on the NIU Chemistry Gopher site) is shown in Figure 5.4. The file types and the MIME types will be discussed later.

All files and subdirectories are added to the .cache file, except those whose names begin with a period. This system is the only means of hiding files from the clients.

Very long menus are difficult to work with, so some thought should be given to establishing a directory tree structure that organizes

Figure 5.4. A .cache file for Gopher.

```
0About the NIU Chemistry Gopher 0/About the NIU Chemistry Gopher
      hackberry.chem.niu.edu 70        text/plain
1Academic Chemistry Employment Clearinghouse  1/ChemJob
      hackberry.chem.niu.edu 70        text/html
1Chemistry Department Graduate Programs 1/Chemistry Department Graduate Programs
      hackberry.chem.niu.edu 70 text/html
1Chemistry Information on the Internet  1/Chemistry Information on
the Internet   hackberry.chem.niu.edu 70 text/html
1Chemistry Software  1/Chemistry Softwarehackberry.chem.niu.edu 70
      text/html
1Conference Listings  1/ConferenceListingshackberry.chem.niu.edu  70
      text/html
1Journal Submission Guidelines  1/Journal Submission Guidelines
      hackberry.chem.niu.edu 70        text/html
1Other Gopher and Information Servers    1/Other Gopher and Information Servers
      gopher.tc.umn.edu 70   gopher_link
1Other Chemistry Gophers      1/Other Chemistry Gophers
      hackberry.chem.niu.edu 70        text/html
1Quantum Chemistry Acronyms Database     1/Quantum Chemistry Acronyms Database
      hackberry.chem.niu.edu 70        text/html
1Stock Prices   1/Stock Prices  hackberry.chem.niu.edu  70
      text/html
0ECCC Information    0/ECCCinformation hackberry.chem.niu.edu 70
      text/plain
```

the files into subgroups that will make it easier for a user to navigate. Most operating systems place no limit on the number of subdirectories in any directory, the level of subdirectories, or the number of files in any directory.

To begin placing data on your server, create the subdirectories you desire into the `root` directory and then copy or create the text files into the appropriate location. The `.cache` files are created automatically, so you can add or delete new files and directories as your server grows.

The file name will be used as the name placed on the menu when the user connects to your server. The simplest procedure is to thus use descriptive file names. However, some systems place serious restrictions on file-name size. Fortunately, a procedure exists for using names in the menu that are different from the file's actual name.

Creating Links. Links allow the Gopher administrator to create connections to a variety of other servers and utilities, and also modify the information in the `.cache` file. Links are created by preparing link files in the appropriate locations and formats. Link files contain data that adds or modifies entries in the `.cache` file. Because they contain no information of use to a client, they must begin with a period to be hidden from view. A single link file can contain a number of different links. Each link is defined by five lines that contain the file characteristics. These five lines are

```
Name=
Type=
Path=
Host=
Port=
```

Each line begins with a keyword, then an equals sign, followed by information. The first line, the `Name` line, contains the name to be given on the menu. This name does not have to be the file name, and in fact, allows for the use of any menu description desired. The `Type` line defines the class of document. The standard gopher server can serve the following types of documents:

- 0, text file
- 1, directory
- 2, electronic phone book server
- 7, full-text index (WAIS)
- 8, Telnet session

(Other numbers can also be used, primarily to handle compressed and binary files.) The `Path` line contains the file name, relative to the `root` directory, with the file type and a slash (/) appended in front. The `Host` and `Port` lines contain the hostname and port number of the server, respectively. An example of a link file (called `.FranceLink`) that offers a name in the menu different from the file name is given in Figure 5.5.

Links can be made to other Gopher servers, simply by having the appropriate link files, particularly the `Host` and `Path` entries. To add a link to the conference listings (which is a directory, requiring `Type=1`) on the NIU Chemistry Gopher site, the link file shown in Figure 5.6 is needed.

Writing link files by hand is not difficult, but the UNIX Gopher client has a built-in feature to make writing links very easy. Using the client, navigate to the menu item for which you wish to create a link. Instead of pressing the Return key (which would make the connection to that item), press the = key. The client displays the entire link file necessary to specify that item. This file can then be copied to the appropriate location, and the link is established. There are no limits to the number of links that can be built into a server.

Gopher can also link to *CSO* or *ph* phone books. A link file is needed to connect to the appropriate server. Gopher handles these

Figure 5.5. File characteristics of the link file called `.FranceLink`.

```
Name=Chemistry Departments in France
Type=0
Path=0/Departments/France
Host=dummy.server.edu
Port=70
```

Figure 5.6. A link file to the directory `ConferenceListings` on a remote server.

```
Name=Conference Listings
Type=1
Path=1/ConferenceListings
Host=hackberry.chem.niu.edu
Port=70
```

Figure 5.7. A link file for a Telnet session to the University of Arizona Fullerene Database.

```
Name=Fullerene Database at the University of Arizona
Type=8
Path=
Host=sabio.arizona.edu
Port=23
```

servers as `Type=2`, so that the client can handle the appropriate protocols for searching these on-line telephone directories.

In addition, you can have your Gopher server initiate a remote login session (called Telnet). Again, all that is needed is the appropriate link file, with `Type=8` to designate the Telnet session. The client will actually hand off the Telnet session to the local Telnet software, supplying the needed address. An example of a link file is given in Figure 5.7. This link is to the University of Arizona Fullerene Database. One drawback to Telnet sessions is that the user must know the account name and password (if needed) to initiate the connection; this information is *not* supplied by the server.

Linking to a WAIS Server

Gopher can also initiate search sessions. The search facility most often used by Gopher is called WAIS. This utility is widespread in GopherSpace and offers a vital service in file retrieval.

Designing a system to enable users to search a collection of files is a complicated endeavor, particularly if you wish to design the system for widespread general use. The user must be able to access the database simply. Selecting the search terms and parameters must be straightforward. Creation of the database must also be simple. For example, no special formatting of the files should be necessary, and selection of keywords should be nontaxing.

What Is WAIS? WAIS was designed exactly to meet these parameters. First developed at Thinking Machines, WAIS is a client–server searching tool. WAIS will take a collection of text documents, in any format, and index them using every word in the texts. No keyword selection is required; every word in the texts is a keyword. Users connect to the server through a simple client, generally having just one field—the word(s) that the user wishes to search for in the database. The server then locates all documents that have the search term, ranks them in relative order, and returns this ranked list of file names. The user then selects the desired file, and the file is then displayed on the screen.

How To Create WAIS Indexes. One can create a standalone WAIS server and connect to it via a standalone WAIS client. However, the Gopher client was developed to directly interface with WAIS so that the Gopher server can also act as a WAIS server. Users can then access the WAIS index through the normal Gopher client.

However, the Gopher server does not have the WAIS server built in; rather it only has the capabilities of incorporating the WAIS server. To incorporate a WAIS server into your Gopher server, you will need to obtain the WAIS code, compile it, and then recompile your Gopher server. We will briefly highlight some of these steps here, but you should refer to the documentation included with the WAIS and Gopher codes to obtain the latest instructions for compiling these systems.

WAIS exists in a number of different "flavors". The version that appears to have the most support and offers the best avenue for continued upgrades is called `freeWAIS` and is a product of the Clearinghouse for Networked Information Discovery and Retrieval (CNIDR). `freeWAIS` can be obtained via anonymous FTP from a variety of sources listed in Table 5.2. `freeWAIS` is copyrighted, but it is available for free. After uncompressing the file and unbundling the tar file, make sure to read the installation notes for particulars for your computer, and then compile the code.

Table 5.2. Locations of `freeWAIS`

Address	*Directory*
ftp.cnidr.org	/pub/NIDR.tools/freewais
boombox.micro.umn.edu	/pub/gopher/Unix
src.doc.ic.ac.uk	/computing/information-systems/NIDR.tools/freewais
nic.swithc.ch	/mirror/gopher/Unix or /mirror/wais/servers/freeWAIS/freewais
bash.cc.kcio.ac.jp	pub/mirror/gopher/Unix

To include the WAIS services within your gopher, you must now compile the Gopher code including the necessary portions from WAIS. Let's suppose that the top level of your Gopher source code distribution is in the directory `/local/gopher` and the top level of the WAIS distribution is `/local/wais`. Change directories to the top level of your Gopher distribution (`cd /local/gopher`) and create the following symbolic links to the necessary WAIS subdirectories:

```
ln -s /local/wais/ir .
ln -s /local/wais/ui .
ln -s /local/wais/bin .
```

If you have previously compiled a version of Gopher, make sure to type `make clean` first, to remove all of the old object and compiled codes. Then recompile gopher as described earlier. The WAIS capabilities will now be included, and your Gopher server will now also act as a WAIS server.

The next task is to create an index of the files you wish to make available for searching. A simple example is given here; the documents included in the Gopher and WAIS distribution have further details if you require them. As a working example, we will examine how we created a searchable index for all of the conference listings at the NIU Chemistry Gopher site.

We have a directory called `/gopher/data/ConferenceListings`, which appears in the top-level gopher menu as Conference Listings. In this directory, we have three subdirectories, `ACSMeetings`, `SpecialistMeetings`, and `GordonConferences`. In each of these directories are more subdirectories and files containing information on conferences (names, dates, locations, speakers, costs, registration fees, etc.). We wish to allow users to run a search on all of these files, and we want the search option to be listed in the menu displayed after the user selects Conference Listings.

First, we must create an index that WAIS will use to conduct the search. Indexes are created using the *`waisindex`* program created upon compilation of the WAIS code. The *`waisindex`* program will create a number of files that are used by the search engine to locate all matches and deliver the file names to the Gopher server. We must now create these index files and then make them available to the Gopher server.

We begin by creating a new subdirectory in the `/gopher/data/ConferenceListings` directory to store the index files. We call this directory `.index` (the leading period will make it invisible to any clients). The next step is to actually create the index files. We want these files to be in the `.index` subdirectory. The *waisindex* program creates the index files we require. The program needs to know the directory containing the files that are to be indexed. In our case, we want to index not just a single directory, but all the files in all the subdirectories as well. The *waisindex* program will allow for this recursive index by using the `-r` flag. The command we then execute to create the needed index files is

```
waisindex -r /gopher/data/ConferenceListings
```

The last step is to create a menu item for the search. We have placed the index files in a "hidden" directory so that users will not be able to directly download the index files (besides, these files are not standard text files). We need to create a link file in the appropriate location to initiate the WAIS search. In our case, we wish the search to appear in the first Conference Listings menu. So we return to proper directory, in our case `/gopher/data/ConferenceListings`, and create the link file, shown in Figure 5.8, which we will call `.indexLink`. Use `Type=7` to indicate that this will initiate a WAIS search and set the Path variable to the `.index.inv` file created by `waisindex`. You now have an operating WAIS search engine. If you add new files or remove files from your directory, you must rerun the `waisindex` program to create updated index files. At the NIU Chemistry Gopher site, we have written shell scripts to reindex the files every day, making use of the *cron* facility on UNIX computers.

Using WAIS. Searches are initiated by the client. The available searches are indicated by a preceding `<S>` on the PC client, a trailing `<?>` on the UNIX client, and an icon of a question mark on the Macintosh client. Selecting a search item brings a search window to the screen. This window contains an entry field where the user can type in any search term he or she desires. Because WAIS indexes every word in

Figure 5.8. A link file to initiate a WAIS search.

```
Name=Search Conference Listings
Type=7
Path=7/ConferenceListings/.index/index.inv
Host=hackberry.chem.niu.edu
Port=70
```

every file, the user can select any word or words as a search term. However, this option can lead to some ridiculous searches; for example, using the search term "the" will probably not be useful.

After the search term has been entered, WAIS finds all files that contain that term and returns an ordered list of file names that contain that term. The files are ranked on the basis of a scheme that tries to determine the files that are most suitable (based on frequency of occurrence and position of term in the file). The best file is given a score of 1000; decreasing scores indicate a less ideal match. If one of the file names looks of interest, all you need do is select that item, and Gopher will deliver that file to your screen.

Some WAIS and Gopher servers can handle Boolean constructs for more refined searching. The details of WAIS searching are outside the scope of this chapter.

The conjunction of WAIS and Gopher is an ideal match. Gopher provides a simple menu-driven document-delivery system, and WAIS provides a simple means for indexing and searching files that Gopher can deliver. This system is not extremely flexible, and the search capabilities are fairly primitive, but all the resources are available without charge and are easy to configure. With some care in the organization of files on the Gopher, a very useful Internet document search-and-delivery system can be initiated and maintained.

Gopher+

The initial Gopher system has a limited set of types of documents that it could deliver. With the development and implementation of MIMEs, the Gopher system needed to expand. We will not discuss MIMEs in detail here, deferring this discussion to Chapter 6. Essentially MIMEs are a description of the file type. To enable handling of MIMEs, the Gopher+ system was developed. The basic operation of the Gopher+ server is identical to the original Gopher server, except that Gopher+ will deliver the appropriate MIME with the file. The client can then interpret the MIME and spawn off the appropriate helper program to handle the file. For example, if a color image, prepared in the JPEG format (discussed in the Glossary), is sent by the Gopher+ server, the receiving client will interpret the accompanying MIME (in this case image/JPEG) and automatically start a graphics program that will display the image on the screen. Gopher+ servers can handle a wide variety of image, audio, and video formats. The current version of Gopher+, available from the University of Minnesota, can handle many MIME types.

Gopher+ implements a +VIEW feature that allows the user to select the desired type of presentation of a file, using the MIME type. For example, a graphic image may be available in PICT, GIF, and TIFF format; the +VIEW feature allows the user to select the desired format. (These formats are discussed in the Glossary).

In addition, Gopher+ will handle ASK blocks, which is a limited form entry procedure. The +ADMIN feature adds an additional tag to the delivered files. This tag can include information on the system administrator. The +ABSTRACT tag contains an abstract of the information in the document. This feature allows for better search capabilities of gopher sites using a program like `Veronica`. `Veronica` essentially indexes all known Gopher sites, allowing a user to search across sites for the information they want. `Veronica` is the Gopher analogue of `Archie`, which indexes known anonymous FTP sites. (Archie and Veronica are discussed in the Glossary).

The GN Server

Two new developments in Internet navigation systems led to the development of a new server. First, the development and widespread implementation of the World Wide Web (described in detail in Chapter 6) has surpassed much of the functions of the Gopher system. For example, the Web integrates the use of text and graphics and handles links via HyperText instead of menus. Second, the University of Minnesota decided to charge commercial sites for the use of the Gopher server software. Academic institutions can still obtain the server code without a fee.

The Free Software Society considers commercial software to be anathema, and has created a wide variety of publicly available software, including c compilers, PostScript viewers, and text editors, under the rubric of its copyright policy. Software is made available for free but remains copyrighted and owned. Distribution of the software is encouraged.

With the principle of this policy in mind, and with the idea of answering the Web challenge, a new server package was developed, called GN, principally by John Franks at Northwestern University.

GN is a single-network navigational server that can handle both Gopher and HyperText transport protocol (HTTP). In one system, GN creates a Gopher and Web server, and the software is available to all for free.

Setting Up the GN Server

The server software can be obtained by anonymous FTP from **ftp.acns.nwu.edu**, and the directory /pub/gn contains the latest version. After you have obtained the file, uncompress and untar the file. We will describe the basic procedure for compiling the GN server, but you should read the documentation for the most recent installation notes.

Most of the site-specific information is contained in the file config.h. You will need to modify many of the items in this file to meet your specific needs. In particular, you must set the variable GN_HOSTNAME as the name of your computer, ROOT_DIR as the complete path name of your root data directory, DEFAULTPORT to the TCP port you will communicate through (usually 70), and USERID as the account number that will run the server (for security reasons this should not be root). Next, check the makefile for any local changes that might be needed. If you wish to interface the server with a WAIS server, just like we described earlier, read the documentation on waisgn. Essentially identical operations need to be done to enable GN to communicate with WAIS as we described for connecting Gopher to WAIS. When all of the configuration variables have been set, compile the program using the make command. This command creates a number of programs: sgn, the standalone version of GN; gn, the version run using inetd; and two utility programs called mkcache and uncache.

GN can be run in two separate ways. The first (and simplest) is to run it as a standalone program. This is accomplished with the command sgn -p *port /path/to/root* &. If the port number is 70 (the default) then this command must be executed by the superuser. In this case, it is very important that the USERID variable in config.h be set to someone without root privileges, in order to maintain as much security as possible. As with Gopher, this command should be placed in a system startup file, like the one shown in Figure 5.3.

GN can also be run using the inetd facility. To do this you will need to add the following line to the TCP services file, usually found as /etc/services (this example uses the default port number):

```
gn  70/tcp
```

Next, edit the `/etc/initd.conf` file to include the following line:

```
gn stream tcp nowait nobody /full/path/for/gn gn
```

Any options for GN (discussed later) can be placed after the last `gn`. The use of the variable `nobody` is to designate a user who does not have `root` privileges. Once this is done, you need to restart the `inetd` daemon. You should consult your local manuals or system administrator for help with this.

A number of options can be selected when starting the GN server. The principal useful option is to enable logging of the use of the server. This is done with the command `-L` *`logfilename`*.

Adding Files to the GN Server: How It Differs from Gopher

Besides the ability to act as a World Wide Web server, GN differs from Gopher in the way files are made available to clients. The procedure is more difficult than with Gopher, but GN allows for more control and greater security of the files and computer.

As with Gopher, GN uses the `.cache` file in each directory to determine what files (and their types) are made available. Gopher creates the `.cache` file automatically; in GN, the administrator must create the `.cache` file. The `.cache` file has an unwieldy format, and so the utility `mkcache` streamlines the process. As input to `mkcache`, the administrator creates a `menu` file that looks very much like the link files of Gopher. A portion of the `menu` file for the top level of the NIU Chemistry Gopher is shown in Figure 5.9.

The five lines that make up the Link file used by Gopher are included for each entry in the `menu` file, along with a few notable additions. The first is the Attribute line, which can be used to define special properties of the entry. In the example, Attribute=invisible, which prohibits this file (or directory) from being listed on any menu to a client. This feature is very useful for it allows files to be placed in convenient locations, but kept them secure, that is, with limited access to them. Files or directories that are not listed at all in the menu file will not be served to any type of client. The invisible file allows for these files to be served under the appropriate conditions, such as by reference within a HyperText document. The Attribute line can also be used to limit entries to be supplied to gopher clients only or to Web clients only.

Figure 5.9. A portion of the `menu` file of the NIU GN server.

```
Name=About the NIU Chemistry Gopher
Path=0/About the NIU Chemistry Gopher
Type=0
Host=hackberry.chem.niu.edu
Port=70

Name=Chemistry Information on the Internet
Path=1/Chemistry Information on the Internet
Type=1
Host=hackberry.chem.niu.edu
Port=70

Name=Conference Listings
Path=1/ConferenceListings
Type=1
Host=hackberry.chem.niu.edu
Port=70

Name=Other Gopher and Information Servers
Path=1/Other Gopher and Information Servers
Type=1+
Host=gopher.tc.umn.edu
Port=70
URL: gopher://gopher.tc.umn.edu:70/11/Other Gopher and Information
Servers

Name=Web Page
Path=0/webpage.html
Attribute=invisible

Name=Chemistry Info Page
Path=0/cheminf.html
Attribute=invisible

Name=Images directory
Path=1/images
Type=1
Attribute=invisible

Name=ECCC directory
Path=1/ECCC
Type=1
Attribute=invisible

Name=Search Chemistry Email Addresses
Path=7g/ChemDir

Name=Electronic Computational Chemistry Conference Information
Path=0h/ECCCinformation
```

If you are converting a gopher server to a GN server, you can readily convert your old .cache file into a menu file by using the uncache program. Typing the command uncache converts the .cache file in the current directory into a menu file. You can then edit the menu file, adding, deleting, or modifying the entries.

Once the menu file has been created, executing the mkcache command creates the properly formatted .cache file that GN will use to serve its menu items.

GN can deliver a wide variety of file types, including graphics, audio, HyperText, and executable files. Just like Gopher it can initiate Telnet sessions and WAIS searches. GN offers an alternative to WAIS for searching a small collection of files.

GN includes a search capability similar to the grep utility within UNIX. To differentiate between these two types of searches, GN has a modified Type designation 7w for a WAIS search and 7g for a grep-like search. The procedure to set up a WAIS search in GN is very similar to that already described for a WAIS search in Gopher. The only difference is the use of the Type=7w for the menu item to select the search.

Creating the grep-like search in GN is very simple. An example is the Chemists' E-mail/Phone Book located on the NIU Chemistry Gopher site. The Phone Book is a collection of files, each one containing information provided by an individual chemist. The files are in the directory /gopher/data/ChemDir, and /gopher/data is the root directory. We need to make two changes to the menu file located in the root directory to make a grep-like search of the phone book publicly available. First, we need to make the entire directory searchable, which is shown in the first entry of Figure 5.10. Using Type=1s makes the files in the directory searchable. Second, we need to create the actual menu item that a user will select. This is done in the second half of Figure 5.10. The ChemDir directory is hidden from the user. The only menu item a client will see is the search entry.

GN can also act as a World Wide Web server, following the HTTP protocols. The details of the World Wide Web are left for Chapter 6 and will not be discussed here. However, GN is a full-featured HTTP server that supports ISMAPS and Forms (discussed in Chapter 6). If you plan on operating both a Gopher and a Web server, GN can provide both services in one easy-to-use package.

Figure 5.10. Setup for searchable directories in GN.

```
Name=Chemists Email Directory
Path=1s/ChemDir
Type=1s
Host=hackberry.chem.niu.edu
Port=70
Attribute=invisible

Name=Search Chemistry Email Addresses
Path=7g/ChemDir
Type=7g
Host=hackberry.chem.niu.edu
Port=70
```

Figure 5.11. A link file for the University of Minnesota list of gopher sites.

```
Name=Gopher Server index
Type=1
Path=1/Other Gopher and Information Servers
Host=gopher.tc.umn.edu
Port=70
```

Maintaining the Server

Registering Your Site

Once you have an operational Gopher server, you need to advertise its existence. The University of Minnesota keeps an extensive list of Gopher servers around the world. To register with this list, send an e-mail message to **gopher@boombox.umn.edu** (to **gopher@ebone.net** if you are in Europe) with the following information: the server's name, the hostname, the port number, and an administrative contact. This list can be obtained through Gopher with the link given in Figure 5.11.

Discussion Groups

A number of discussion groups are devoted to Gopher. Discussion groups are an excellent means for learning of system updates and code patches. In addition, they are a perfect means for asking questions and obtaining answers from users worldwide. The Usenet for Gopher is **comp.infosystems.gopher**. A mailing list devoted to Gopher is operated by the Gopher developers. To subscribe to this list, send a message to **gopher-news-request@boombox.micro.umn.edu**. A discussion list devoted to GN is also available. To subscribe to this list, send the message `subscribe gn-maint-l` to **majordomo@hmc.edu**.

Monitoring Usage

Keeping track of the use of your Gopher server can help to identify features that are useful and those that are being neglected. This information can lead you toward developing new products and services. A number of different utilities are available for monitoring the use of the Gopher server. All of these utilities require that the logging capability be turned on. The basic features of these tracking programs are similar. They prepare compilations of the files and directories that were accessed, the clients who have accessed your server, and the number of times each file was selected. They also provide graphical output of the results. We present a list of a number of these programs and the location for obtaining them in Table 5.3. The choice is very much a personal one, based on what presentation features and options best match your desires.

Future Developments

The continued growth of Gopher within the chemical community is dependent on two factors. First, the Gopher concept is perhaps limiting. The World Wide Web offers essentially all the features of Gopher

Table 5.3. Locations of Gopher Tracking Programs

Program	*Location*
GETSTATS	anon FTP to **ftp.eit.com**; obtain latest version in the directory `/pub/web.software/getstats`
GNLOG	included in the GN distribution
WUSAGE	anon FTP to **isis.cshl.org**; obtain latest version in the directory `/pub/wusage`
GLOG	anon FTP to **boombox.micro.umn.edu**; obtain latest version in the directory `/pub/gopher/Unix/GopherTools/glog`

with the added benefit of HyperText and inline graphics. Certainly, some of these features can be added to Gopher, but the explosive growth in Web traffic suggests that it is the technology of the future. However, Gopher has the advantage of a very simple interface and a simple server setup.

Second, and more importantly, a Gopher server is only as good as the information it holds. Many chemistry-related Gopher sites contain pointers only to other Gopher servers. This approach may provide some service to the local community (for example, a chemistry department) but holds no interest for other chemists. Chemists will begin to make use of the Internet, Gopher, and the World Wide Web only when they find useful information and tools in GopherSpace and the Web. Fancy links and graphics will not increase the number of Gopher users. Future Gopher (and WWW) administrators must recognize the need for information first.

chapter six

The World Wide Web

Henry S. Rzepa

In This Chapter

This chapter introduces the World Wide Web and explains, using chemical information as an example, how it provides a means of processing, disseminating, and perhaps even increasing the quality of available information. This chapter is essentially a starter kit for those unfamiliar with the World Wide Web, illustrating how to navigate the system and how to develop original material for it. (A glossary of the terms used in this chapter is given in Table 6.1.)

3223–7/96/0205$37.75/1

Table 6.1. Glossary of Terms Associated with the World Wide Web

Term	*Definition*
Bookmark	A user pointer to a previously visited World Wide Web document
Browser	A client program that displays World Wide Web documents
CCI	Common client interface
CERN	European Laboratory for Particle Physics
CGI	Common gateway interface
CML	Chemical markup language
CSML	Chemical structure markup language
DNS	Domain Name Server
DTD	Document-type definition
Firewall	A mechanism for providing secure access to and from the Internet
FORM	A graphical interface for interacting with World Wide Web services
GIF	Graphics interchange format
Helper	An independent program invoked from a World Wide Web browser
HTML	HyperText markup language
HTTP	HyperText transport protocol
Hyperglossary	A glossary of terms created and accessed via the World Wide Web
Hyperlink	A pointer within a document to another network object
Inline	An image inserted within text and displayed by a World Wide Web browser
IP	Internet protocol
JPEG	Joint Photographic Experts Group (image format)
MIME	Multipurpose Internet mail extension
MPEG	Moving Picture Experts Group (video animation format)
NCSA	National Center for Supercomputing Applications
OpenDoc	Compound document architecture system
PHP	Personal home page on the World Wide Web
PNG	Portable network graphics format
PPP	Point-to-point protocol
PROXY	An alternative nominated HTTP server
RTF	Rich-text-format (or Microsoft interchange format)
Server	A program that provides access to a database of World Wide Web documents
SGML	Standard generalized markup language
SLIP	Serial-line Internet protocol
SSL	Secure-sockets-layer transport protocol
URC	Uniform resource characteristic
URL	Uniform resource locator
URN	Uniform resource name
VRML	Virtual reality modeling language
W3O	World Wide Web Organization
WAIS	Wide Area Information Service
Web	World Wide Web
Webmaster	The administrator of a World Wide Web server
WWW	World Wide Web

THE WORLD WIDE WEB is the branch of the Internet that consists of a system of interlinked documents that may include text, graphics, sound, movies, and links to still other documents and services. The World Wide Web is thought by many to represent the first coherent unifying mechanism for users of the Internet. A measure of the popularity of this system is that the number of chemically oriented World Wide Web servers increased from less than 10 at the start of 1994 to more than 150 in March 1995.

HyperText and Hypermedia

In a conventional book or document, the content flows in a serial fashion, and references and citations are flagged and then postponed to the end of the section or chapter. Alan Kay[1] foresaw that computer technology could enable the textual content of a book, or more generally hyperdocument, to be much more actively cross-linked by means of a device known as a "hyperlink", whereby the order of information presented to the users is much more under their control and is not determined by the physical pagination and binding of paper. With significant foresight, Kay suggested that the hyperlink could itself contain some form of algorithm that would act upon information entered by the user. The hyperdocument could in effect respond dynamically to the user and present information unique to any particular session. Kay is generally credited with introducing the term "Dynabook" or "dynamic book".

Many of Kay's ideas were developed into a product called Hypercard for Apple Macintosh computers in the mid-1980s and have subsequently been followed by a number of other commercial tools and products, including a delightful range of "Living Books" in which almost all the text has been replaced by animated graphics. With the exception of some teaching applications, the concept has not found widespread acceptance among the chemical community as a publishing medium. Part of the reason is probably that the authoring of such works requires learning arcane and often nonstandard or proprietary scripting languages.

Conventional HyperText documents were also limited in their utility because all of the information had to be physically present on the data-storage system of the computer being used. Chemical information by its very nature has evolved into a variety of modular and distributed data sources, and the task of consolidating any particular subject or theme into a single location is not a trivial one, as the activities of the Chemical Abstracts Service illustrate.

Within a paradigm that accepts the diaspora of chemical information on a global scale, we need a mechanism to uniquely identify the resources or objects, and link them using a suitable computer network such that they can be accessed within seconds. The object in this context can mean a text-based document or data file, a graphic image, sound, animation, or more generally a set of instructions for performing further actions. To facilitate user acceptance, it is also essential that most of the protocol for acquiring and presenting the data be hidden behind the HyperText concept in a suitably transparent manner.

The World Wide Web System

In 1989, Tim Berners-Lee[2] working at CERN, the European Laboratory for Particle Physics in Geneva, Switzerland, introduced a very simple and elegant extension to the original Dynabook concept. This extension involved recognizing that the global Internet system represented a highly homogenous and widespread communications infrastructure that would allow transparent transfer of information from one physical location to another. In effect, the entire Internet could be used to create a single large Dynabook of information.

The second breakthrough resulted from the development of SGML, or standard generalized markup language,[3] and its ratification as an international standard. In the past few years SGML has been widely accepted by the scientific publishing community. This development has enabled the creation of a diverse set of formal document-type definitions (DTDs) that describe how complex technical documents can be "marked-up" in a digital sense. One DTD, (ISO 12083[4]), was designed specifically for scholarly scientific publishing.

With these two supports in place, the scene was set to create a nonproprietary and global communication system called the World Wide Web, abbreviated variously as WWW, W3, or just "the Web". Information is uniquely and globally identified within this system, currently using labels known as uniform resource locators or URLs.[5] Ultimately, every unit of consolidated chemical information might eventually be associated with such a URL, and the challenge becomes one of organizing and indexing information within this metaphor.[6]

HyperText Transport Protocol (HTTP)

A short preamble to the two principal components of the World Wide Web is appropriate at this stage to introduce the quite distinct roles they play in the overall mechanism.

The first component is known as HTTP[7] or HyperText transport protocol. As will become apparent later, this has become something of a misnomer in referring specifically to text, because the highly visual content of chemical information is handled equally well. HTTP is something best left as a "black box", and a simple analogy serves to illustrate why. The familiar actions of selecting a print action from a computer and a few seconds later collecting output from a laser printer involve the transmission of a set of instructions such as PostScript from the computer to the printer. In this sense, PostScript is a language intended for communication purely between machines, and is rarely intercepted directly by people. By analogy, HTTP functions to exchange information between a WWW ***server***, which holds information, and a WWW ***browser***, which displays it. One characteristic feature of HTTP is that it is a "stateless" protocol, in that a typical WWW session may consist of many separate HTTP-based transactions between which no mutual "state" or context is maintained.

HTTP was originally defined and developed at CERN. The exponential growth of the WWW during the period 1990–1995 occurred in large measure because CERN and other organizations made the source code freely available in the form of a common code library to anyone wishing to develop applications. HTTP is analogous to other Internet communication protocols, such as Telnet, FTP (file-transfer protocol), and Gopher, which are discussed in other chapters of this book.

HyperText Markup Language (HTML)

The second component of the WWW is known as HTML or HyperText markup language. This symbolism follows the semantic rules defined by the SGML[8–10] standard. HTML files are invariably created as text files, and as such they can be readily edited and interchanged between different computer systems. In this they are analogous to documents saved in the RTF or rich-text format, introduced by Microsoft as a mechanism to enable the exchange of formatted documents. Because of the precise semantic content of HTML, the markup is potentially treatable as a black box to be hidden behind the user-friendly interfaces of editing, word processing, data acquisition, and database programs.

In 1994 a significant proportion of original HTML documents were being written using simple text-editing programs, but this method is expected to become a rare phenomenon as better tools become available for automating the process. For example, software developments such as the `Internet Assistant`[11] for use with Microsoft `Word` or the `Internet Publisher` for `WordPerfect`[12] will mean that the steep learning curve often perceived to be associated with programming languages need not be experienced.

Programs that implement both the HTML page description and the HTTP communication protocols are known as WWW browsers. Programs that support only the HTML definition are known simply as HTML authoring tools, and support of purely the HTTP component is the task of WWW servers. Both protocols are still developing. Version 2.0 of HTML will be superseded in 1995 by version 3.0, and will be complemented by a three-dimensional descriptor known as VRML (virtual reality modeling language). HTTP is thought likely to evolve to the new generation HTTPng. These developments will be reviewed later in this chapter.

Uniform Resource Locators, Names, and Characteristics

The most interesting feature of HTML, which differentiates it from other markup styles, word-processing programs, or Hypercard stacks, is the inclusion of an Internet-aware component known as a uniform resource locator or URL. Typical examples of URLs are as follows

http://www.ch.ic.ac.uk/WWW/
gopher://gopher_server.country/11/WWW
ftp://anonymous_server.country/pub/dummy.tar
telnet://service_machine.country
news:sci.chem
mailto:dummy@institute.country

URLs are almost always case-sensitive and so should be typed exactly as shown.

The structure of the URL can be approximately divided into three components: the communication protocol, followed by a unique Internet address for the computer system supporting this protocol, and optionally a path to a document or a file present on this computer. The protocol definition can be optionally omitted with some browsers such as `Netscape`. Thus the URL **www.ch.ic.ac.uk/WWW/** will default to HTTP, a reasonable assumption given that in terms of the sheer number of Internet transactions, HTTP has now overtaken the older Gopher and FTP transport mechanisms.

In the URL **news:sci.chem**, news refers to the network news-transfer protocol (NNTP) used as the carrier for Usenet Internet discussion groups. In URLs starting with **mailto**, `mailto` uses the simple mail-transfer protocol or SMTP required for exchanging Internet electronic mail. All WWW browsers support HTTP, Gopher, and FTP internally, and most now also support news and `mailto`. Telnet sessions are normally delegated to a separate terminal emulation program.

The weak link in the URL is the explicit association of the computer file containing information with a directory structure and the named machine used for its storage. Computer systems have finite lifetimes, and directory structures often suffer re-organization, either of which could render the URL inoperative. The first versions of a new addressing scheme to be known as uniform resource names (URNs) are currently being formulated.[13] These take the form

urn:dns:mitra.path.net:1a2b3c4d

in which a domain name server (DNS) is specified (mitra.path.net in this example); it serves to "resolve" a globally unique document identifier (for example `1a2b3c4d`) into an explicit file on a defined computer system. The "opaque" identifier will be allocated to the creator of the document upon request by a "naming authority", and it is expected that these names will persist on a timescale of at least decades. If the document, or the computer it resides on, has to be relocated for whatever reason, the administrator simply re-registers the identifier via the naming authority, a process entirely transparent to the reader of the document. A uniform resource characteristic (URC) will be used to identify the cost, size, authorship, date, and other attributes of the document for the reader to assess. Implementation of URNs and URCs will require a global infrastructure to be established, a process that may take some time to complete.

World Wide Web Conferences

The time scales associated with the medium you are currently viewing are substantially mismatched with the pace of evolution of the WWW.

The WWW community has thus relied on a combination of the Web itself to document the detail of the evolving standards, and international conferences to disseminate the general directions being taken. Most recently, the role of coordinating activity in this area has devolved from CERN itself to the specially created World Wide Web Organization (W3O), located at the Massachusetts Institute of Technology in Boston, MA, with a sister arm (EuroW3O) at INRIA (Institut National de Recherche en Informatique et en Automatique) in France. Many footnotes and citations at the end of this chapter appear in the form of URL references rather than to scholarly papers or books, reflecting the pace at which development of the Web has occurred.

The first two global conferences devoted to WWW developments were held during 1994, at CERN in May[14] and in Chicago, IL, in October,[15] and two more conferences are planned for 1995 and 1996. Many of the subjects discussed here were presented for the first time at these meetings. A significant proportion of the administration and registration for these events was also accomplished using the mechanisms of the Web. Smaller regional conferences are also being organized, and in due course subject-specific meetings may also arise. As discussed elsewhere in this chapter, by mid-1995 three WWW-based virtual chemical conferences will have been organized, and it is now common for international meetings and conferences to be advertised via associated Web pages.[16]

World Wide Web Browsers

This section deals with the generic details of how to select and set up a WWW browser program. The first step is to establish if the computer to be used is already "Internet-ready"; that is, having an allocated Internet protocol (IP) address and name and a wide-area connectivity beyond any local network. This setup is likely to be true of a UNIX-based workstation, which is unlikely to require further configuration. Microsoft Windows and Macintosh systems must have specific IP software installed and ideally a permanent IP number and name allocated by a system administrator. The OS/2 Warp, Windows-NT 3.5, and Windows-95 operating systems are also capable of dynamic allocation of IP addresses without explicit intervention by the user, something also true of certain forms of access via Internet gateways. Certain implementations of HTTP servers permit access to documents to be controlled via the explicit IP name or number of the client computer, and therefore it is always advisable to try to allocate a permanent rather than a dynamic IP address.

If an individual computer is not yet IP enabled, it will be necessary to arrange installation of a suitable cable to the nearest suitable hub, router, or gateway. Internet access from home or other locations without a permanent network infrastructure will require a modem or Integrated Services Digital Network (ISDN) adapter with either PPP (point-to-point-protocol) or SLIP (serial-line Internet protocol) connection software and a dial-up account to an Internet gateway supporting such a service. Internet gateways are available from network providers in many countries. Your organization may be behind a mechanism called a security "firewall",[17] which will require specific local arrangements. In the future, the existing IP networking protocols (in 1995 known as IPv4) will gradually be replaced by the new IPv6 protocols.[18] These will offer a much greater addressing range, along with more responsive ways of coping with the enormous increase expected in Internet traffic, but may also require changes to the way that the local Internet infrastructure is implemented.

Acquiring a WWW Browser

A window-based browser offering a graphical user interface is highly recommended. The program generally associated with the early popularity of the Web is the National Center for Supercomputing Applications (NCSA) *Mosaic*, still currently available at no charge for X-Windows, Microsoft Windows, and Macintosh computers. Browsers are also available for computers such as Amiga, operating systems such as DEC VMS or NeXTStep, and display devices such as VT-100 terminals (*Lynx*) or even the Apple Newton Message pad.[19] The latest generation of browsers, such as *Netscape*, offer features that include newsgroup browsing and encrypted transactions, and more specialized programs such as Arena can offer mathematical typesetting and other advanced aspects of the HTML language. The variety of purely commercial products is also expanding, and it can only be a matter of time before a browser specifically tuned to the needs of chemists is produced.

The best way to acquire the very latest version of any particular WWW browser is via anonymous FTP transfer using either a standalone transfer program or a WWW browser itself (if you wish to update your current copy). Details of the FTP connections to various sources of WWW software are given in Table 6.2. You should endeavor to connect to "mirror" sites rather than to primary sites if at all possible to ensure the fastest network transfer speeds.

Subsequent updates to the programs or acquisition of alternative viewers can be accomplished from within the browser by selecting help or home page menu items. More conventional operations are available, such as copying from another machine onto a floppy disk or from a departmental file server where permitted, and suitable browsers should be available in good shareware collections.

Setting Up a WWW Browser

Most browsers should work "straight out of the box" if your Internet connection is active. To gain the full potential of the program, however, you are advised to enter various specific preferences associated with your local environment.

Local Internet Information. Check that your computer is configured to recognize a domain name server (DNS). Almost all URL references in the WWW are in the form of text-based host names such as **www.dept.org.country**. These host names have to be resolved into a numeric form, such as 123.456.7.89, by your local DNS. If a DNS is not active or is not set up correctly, you will have difficulty making any connections, because most hyperlinks are expressed as named rather than numeric addresses. Defining the DNS entry is normally accomplished during the initial network setup of the computer (Figure 6.1).

Table 6.2. Sites Providing FTP Access to WWW Browsers

Internet Address	*Program*	*Directory*
ftp.w3.org	*Arena*	pub/www/arena
ftp.ncsa.uiuc.edu	*Mosaic*	Web/
ftp.ora.com	*Viola for X*	pub/www/viola
ftp.cs.unlv.edu	*Chimera*	pub/chimera
ftp.einet.net	*MacWeb* and *WinWeb*	einet
ftp.mcom.com	*Netscape*	pub/netscape
src.doc.ic.ac.uk	All packages	

Figure 6.1. A DNS setup for a Macintosh computer.

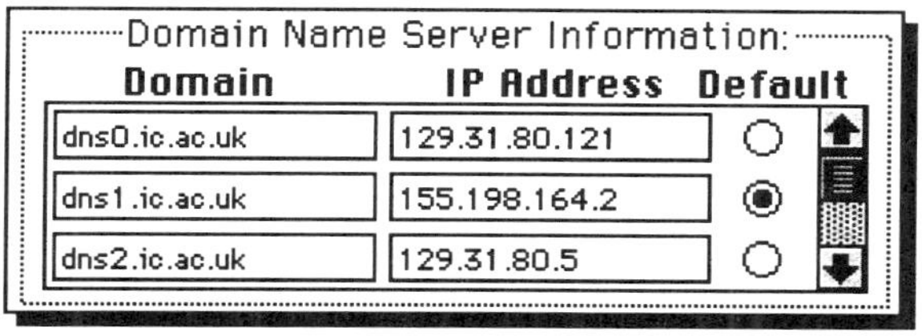

Check also on the name your own computer as registered with the local DNS, because this entry will appear in usage logs of any WWW server you may connect to. As this last statement implies, do not visit any server where you do not want your presence recorded.

Have the name of your local Usenet news reader ready if you want to access such items via your WWW browser. A typical entry for NCSA *Mosaic* for Windows would be

```
NNTP_Server="news.yourhost"
```

Most browsers require entry of your electronic mail (e-mail) address into a preferences or an initialization file.

```
E-mail=yourid@yourmailhost
```

This address is used as the login password in any anonymous FTP transfers that may be invoked during your browsing, and will also be used when any hyperlinks of the form `mailto:name @address` are invoked. If your machine is likely to be used by others, remember that their browsing will be recorded in your name!

You may wish to define your local Wide Area Information Service (WAIS) server to enable global indexed searching. Some servers have been locally WAIS-indexed, and interfaces to WWW browsers such as *wwwwais* may be available to you.

If you are behind a commercial firewall, you may need to connect to the Internet via a proxy server, sometimes referred to as an application gateway or forwarder. This is an application that mediates traffic between a protected network and the Internet. SOCKS is a generic proxy system that can be compiled into a client-side application (implemented by the browser rather than the server) to make it work through a firewall.

Figure 6.2. Configuring local support services using *Netscape*.

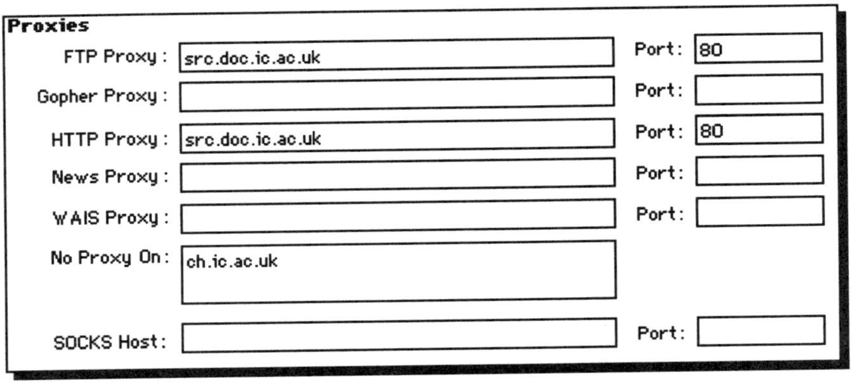

Your organization may be using a so-called "caching" server, through which all local requests are funneled. Use of such servers may require defining a proxy connection in your local browser. Support for such proxy services is becoming more popular as the congestion on the Internet increases, and indeed may become mandatory in the near future. A typical setup display for such services is shown in Figure 6.2.

The Browser Home Page and Other Browser Defaults. When a browser is first started, a default HTML document known as the home page can be automatically loaded. A pristine WWW browser with default preferences will probably have the home page pointing to the organization responsible for the program and identified by a suitable URL. You may wish to change that default to your own personal home page (PHP) or to that of your department or institute, or you may chose not to have one defined at all. Your choice should be determined by whatever home page has the most appropriate hyperlinks defined for your interests and requirements, and also whether your computer is permanently connected to the Internet or not. Home pages defined as pointing to an HTML document on the local hard disk need not invoke any network connection, and those pointing to a very large remote document will probably mean a long startup period as the document is transferred. Most servers also implement a convention that a document with the name `.index.html`, `index.html`, or `welcome.html` in a given directory is invoked if no explicit file name is given. Two de facto global index pages for the World Wide Web are suitable starting points for developing your own pointers:

```
http://www.w3.org/
http://www.ncsa.uiuc.edu/SDG/Software/Mosaic/Docs/web-
index.html
```

The URL of the home page must be entered into the browser configuration, and unfortunately there is no common way for doing this. (A URL is case-sensitive and must be typed exactly as shown.) On the X-Windows-based *Mosaic* browsers, the URL of the home page is entered globally by adding the following entry to an X-resources file called `Mosaic` located in the directory `/usr/lib/X11/app-defaults:`

```
Mosaic*homeDocument: http://server-name/homepagename.html
```

Alternatively, the same entry added to a file called `.Xdefaults` placed in your personal home directory will apply only to your own UNIX account. In Macintosh browsers, there is usually a Preferences menu item suitable for the task. On MS Windows *Mosaic*, the entry is configured into a file called `Mosaic.ini`

```
Autoload Home Page=yes
Home Page=http://server-name/homepagename.html
```

The *Netscape* browser offers a graphical user interface for the creation of the preferences file. Any default home URL that refers to the HTTP protocol will require an active network connection, a possible inconvenience if you wish to use the browser to display only local documents. In this case, a URL of the form pointing purely to a local disk file can be used:

```
file:///hard_disk_name/directory/filename.html
```

Client and Author Style Sheets. Another feature of the WWW system is that to a significant extent the "look-and-feel" of the browser display is under the control of the user, in the form of style specifications. These deal with the type, size, color, and attributes of the fonts that will be used to display text (Figure 6.3). Such flexibility means that a WWW browser can be used in one extreme as presentation software for projection onto screens, with large (>24 point) font sizes, with, for example, white text on a black background and yellow and red colored hyperlinks, and at the other extreme for printing to a monochrome device in a small typeface. This very flexibility has also caused some concern to publishers, to whom the look-and-feel of a document can be of critical importance. This concern will be addressed in the future

Figure 6.3. Setting styles in the *Netscape* browser.

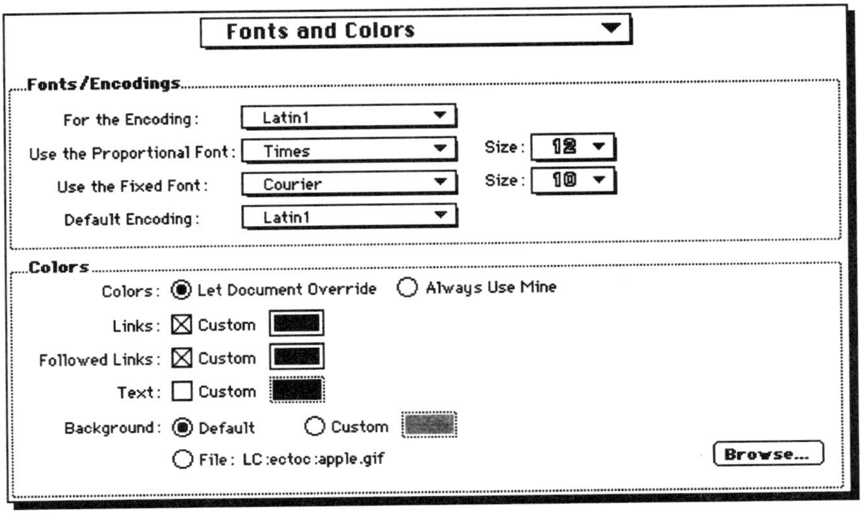

by the implementation of downloadable style sheets[20] containing the document author's preferences for the display. This proposed implementation means that within the browser, the user will still be able to select a weight for each style attribute ranging from 0 (user controls style) to 1 (author controls style).

Using a WWW Browser in a Minimal Configuration. At this stage, you are ready to start exploring the Internet Web. The home page you have selected should ideally have one or more embedded hyperlinks to other sources of information. Such hyperlinks can be represented in at least four different ways on the screen (Figure 6.4). Hyperlinks embedded in text are normally shown both in color and optionally as underlined text. The URL corresponding to a hyperlink can be seen in a small dialogue box in the browser window if the cursor is moved across the text. On some browsers, a prolonged mouse press on a hyperlink can enable further operations to be performed (Figure 6.5). When a link is activated by a single mouse click, most browsers show a status indicator of the progress of the data transfer, and, when complete, the contents of the browser window are replaced with whatever document the link was associated with.

Figure 6.4. Typical examples of hyperlinks.

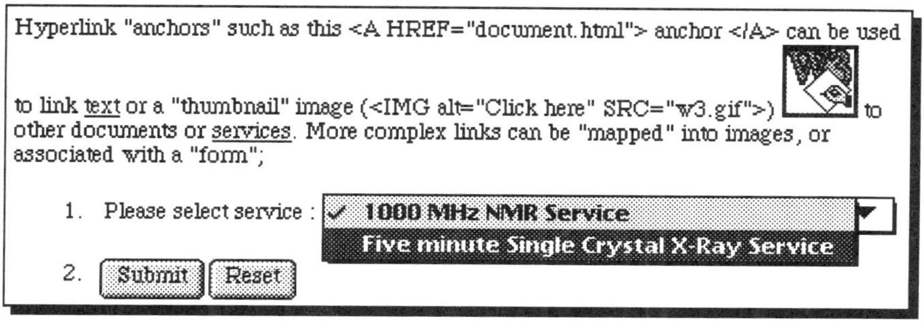

Figure 6.5. Optional actions possible on a hyperlink entry.

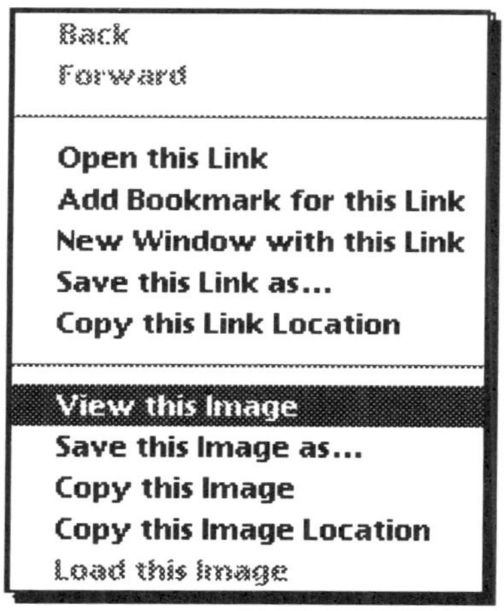

Another type of hyperlink can be represented by a small colored icon, the presence of which implies that an "in-lined" graphical image can be presented within the existing text document. This ability to send images was not a feature of the very earliest HTTP servers, which explains the presence of the word "text" in HyperText transport protocol. Activating such links should be done with an awareness that the image can be relatively large (>100 kbytes) and that this may take some time to transfer and display. For this reason, the user is always given a choice in the setup preferences of whether all such images should be automatically acquired or whether they should all be left unexpanded. In the unexpanded state, the ALT="text" entry is displayed by some browsers to indicate at least approximately the nature of the image. A good default option is not to automatically expand images, because they can be individually expanded later as desired. Some browsers, such as `Netscape`, allow the acquisition of images to occur concurrently with other hyperlink expansions, and the image itself can be interlaced to more rapidly achieve a recognizable picture.

An image surrounded by a colored border indicates that the image itself is a hyperlink to another document. Individual regions within an image can also be used as hyperlinks, a technique known as image mapping. The details of how to apply and implement image mapping will be discussed later. A more subtle form of link is presented in the form of pull-down menus and submit buttons that can enable quite complex actions to be invoked. Such Forms descriptors are also elaborated later in this chapter.

Document Navigational Aids. During the course of a session, many documents and images may be inspected, and some means of navigating around these is necessary. All browsers implement at least three methods for navigating. First, a small document toolbar allows the user to find the home page again, to go one back or one forward in the pages visited, or to reload the current page or images (Figure 6.6).

Second, the browser keeps a list of the titles of each of the documents displayed in a history file, and these can normally be accessed from a menu item. The history file, however, represents only the current session, and the user can chose to make individual entries in the history file more permanent by adding them to a hot list or a bookmark file using the appropriate pull-down menu item of the browser. This file is saved to disk and will be available for future sessions. Various

Figure 6.6. Document navigation toolbars.

bookmark files corresponding to different needs can be saved and passed on to other users, and utility programs exist for converting a bookmark file into a marked-up HTML document.

Despite these aids, the user can easily end up disorientated. One potential navigation aid, which is unfortunately rarely implemented, is the so-called "link" element in the HTML language. The link element could provide a means for an author to indicate a static tree structure for a document by internally pointing to a parent, next, or previous item. This step would facilitate tasks such as creating slide shows, automated printing of complex collections of documents, and maintaining a version history. Another solution to the navigational problem, as implemented in the Hyper-G system[21] developed at Graz in Austria, is to provide an on-screen display of the visited document tree. Other proposals include extensions to the HTML language to support document-specific toolbars and author-predefined guided tours. Until a generally accepted mechanism is identified, the best solution is for authors to build navigational aids into their own material, such as, for example, inserting hyperlinks relating a document to the home page of any "document cluster".

Network Bandwidth and Local Mode. In theory, you need never know which part of the globe any document or information is coming from. In reality you will soon notice that some documents are retrieved rather more quickly than others! For example, if you are connected via a 10–100-Mbit/second Ethernet link to a departmental or institutional server, the transfer should occur at speeds of 10 kbytes/second and greater. Transfers between continents may occur well below 1 kbyte/second, depending on the time of day. Users soon learn to avoid certain types of file, particularly video animation files in MPEG (motion picture experts group) format, which can be larger than 1 Mbyte.

The current solution to network congestion is to set up a so-called "caching" HTTP server, in which a large disk array is devoted to storing all incoming requested documents. This setup does nothing to improve the speed of transfer for the first request for any document, but subsequent retrievals are at local network speeds. The files then "age" and will eventually be purged if not requested again. The disadvantage of such a system is that the user may acquire a document that is no longer current. To counter this disadvantage, authors of documents will be able to specify an expiration period, after which a new copy must be retrieved from the remote source rather than from the caching server. Such solutions may only postpone rather than solve the problem, and we may be in for a period of frustrating delays in downloading files from remote sources around the world.

Most browsers have a prominent stop button to abort a transfer that is clearly going to take an inconvenient length of time. A feature not yet universally supported is allowing the browser to indicate the size of a remote document before retrieval is attempted by requesting header information from the HTTP server. This process requires only a few bytes of data to be transferred initially, and allows the user to decide whether to proceed or not. If the network connection is either down or performing slowly for any reason, this does not prevent documents from being retrieved from a local hard disk or CD-ROM. When necessary, updates to such local information can be acquired using the network connection. Such a model serves well for a teaching reference work in which the bulk of the material is not going to change but needs to be accessed rapidly and reliably. Course updates, set problems with model answers, and other topical material would be acquired from the Internet, although the user need not be aware of the difference. In this spirit, this chapter of the book is associated with a home page, **http://www.ch.ic.ac.uk/WWW/**, containing specific updates to information that may no longer be current by the time this book has appeared in print.

User Input via Browsers

WWW browsing as discussed thus far is essentially a read-only mechanism for information flow. In fact, the original concept described by Berners-Lee[2] involved an active two-way read–write collaborative mechanism, a concept recently revived with the announcement of the Symposia editor–browser package.[22] Perhaps the failure of the Web to develop rapidly in this direction might be attributed to the notoriety of virus propagation during the early 1990s and the lack of use of suitably secure mechanisms on the Internet. In any event, authoring and mounting Web documents has to a large extent developed as a separate area, requiring specific software tools, and with responsibilities normally delegated to a formal "Webmaster" and/or "Docmaster". Discussion of such Web authoring is deferred until later in this chapter. Several secure mechanisms for interacting with and contributing to information content on the Web have nevertheless evolved, and these are elaborated now.

Annotation. Some browsers such as NCSA `Mosaic` allow a local annotation to a document to be created, either as a text-only addition or as a sound file. These additions are stored on the local hard disk and are not available to others, although they can be permanently incorporated into documents if desired. This feature has not been widely adopted in other browser systems.

Newsgroups and Electronic Mail. Usenet newsgroups can be accessed directly from within a WWW browser (Figure 6.7). Individual postings in any newsgroup category can be browsed via hyperlinks, and an e-mail response can be formulated (Figure 6.8).

Electronic messages can be also sent directly by invoking the `mailto:` command as a URL entry (Figure 6.9). Also frequently supported is the ability to enclose documents as e-mail attachments.

Interactive Forms Entry. It is not normally possible to modify remotely stored documents during the process of browsing. The one exception to this situation is via an interface known as Forms entry.

Figure 6.7. Configuring newsgroups within a WWW browser.

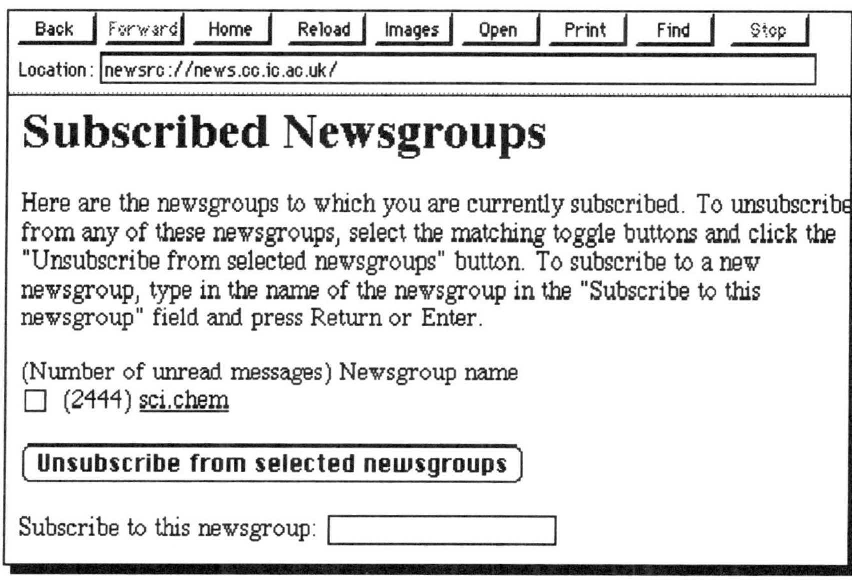

Figure 6.8. Interaction with the proceedings of a newsgroup via a WWW browser.

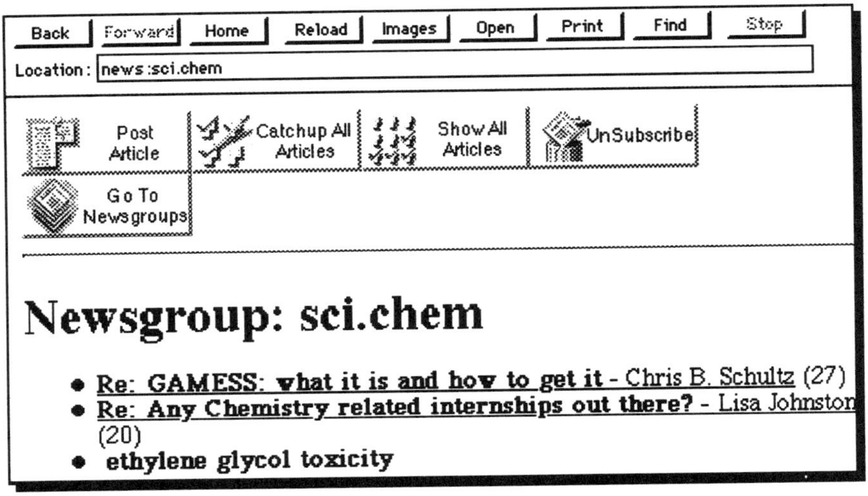

Figure 6.9. The e-mail interface in a WWW browser.

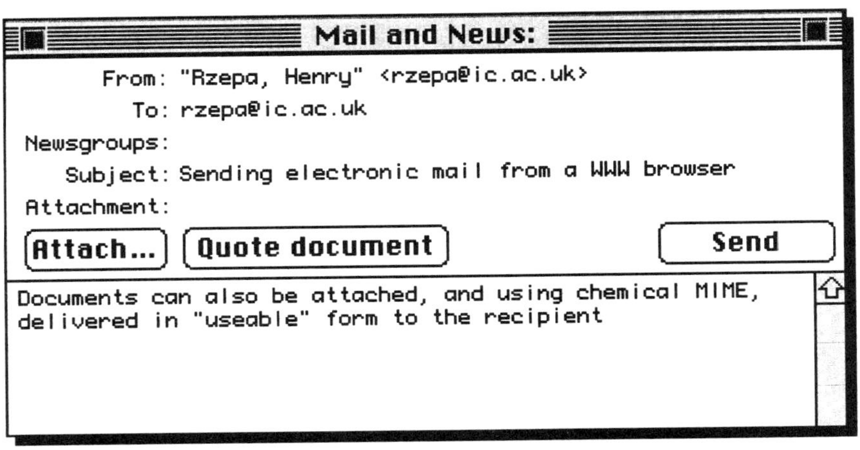

This is possible because only well-defined response fields are available and the user entries in these fields are processed at the WWW server according to a predetermined formula and in a secure fashion. Typically, Forms have been used to register participants at conferences or to allow remote support queries to be submitted to a service center. Such interaction opens entirely new vistas for communication, such as a teaching course in which Forms are used to enable students on the course to respond to set problems, or for authors of papers or textbooks to solicit responses from their readers. In theory, these responses could be collected and automatically added to the content of the original document. An example of form-based input can be found on the this chapter's home page (**http://www.ch.ic.ac.uk/WWW/**).

Another development is the use of such Forms to allow the submission of molecular data[23] that can be used to create a remote database or hyperglossary (Figure 6.10).[24]

In the background, scripts or programs can be used to verify the integrity of the data received, and if necessary to transform it into the

Figure 6.10. A Forms-based entry for a molecular hyperglossary.

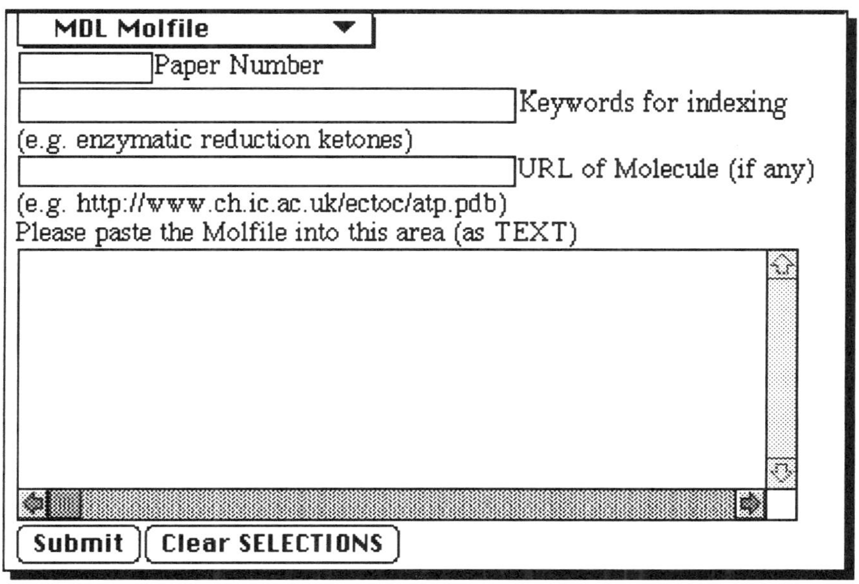

appropriate format for the database. Such mechanisms could also be used, for example, to solicit contributions at conferences or to electronic journals or for refereeing of submitted articles.

Interactive Keyword Searching. It is becoming common for browsers to allow keyword searches based on the text content of the document being displayed. A hyperlink can also be defined to a remote index search engine, which allows simple text and logical search operators to be passed to a remote program. If an HTML entry of the following type

```
<ISINDEX ACTION="http://machine/cgi-bin/search-engine">
```

is present in a document, this will normally be indicated in the display (Figure 6.11).

The search program will normally return any results back for display in the browser window. Such searches are performed using, for example, the WAIS or Lycos system.[25] A collection of various search engines is available in which the queries are actually submitted via a Forms interface (Figure 6.12).

Search and indexing systems such as WAIS or Lycos are based on the text content of a document, whereas other types of search string such as a molecular formula can be passed to a remote script (Figure 6.13).[26] This script can then invoke other programs that perform some action on the molecular formula, mark the results in either graphical or HTML format, and have the answer displayed back on the browser (Figure 6.14). Such a mechanism has been used to calculate isotope mass ratios for mass spectroscopic requirements, although it is easy to imagine how such a procedure can be applied to many other areas of chemistry.

Molecular structure searches based on similar entry of molecule description strings in SMILES[27] or other formats are also entirely feasible, and use of this feature is also facilitated by the support of cut-and-paste mechanisms from other programs such as word processors

Figure 6.11. A link to an index entry.

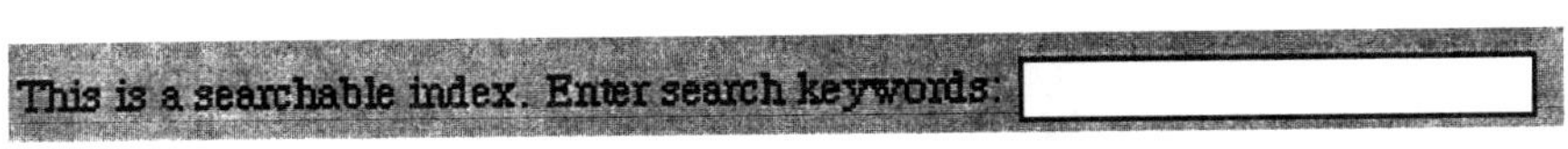

Figure 6.12. A collection of global search engines.

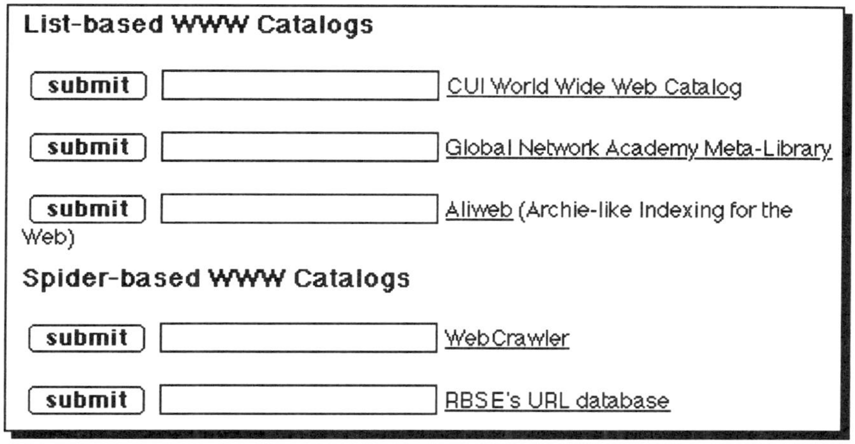

Figure 6.13. A search entry corresponding to a molecular formula.

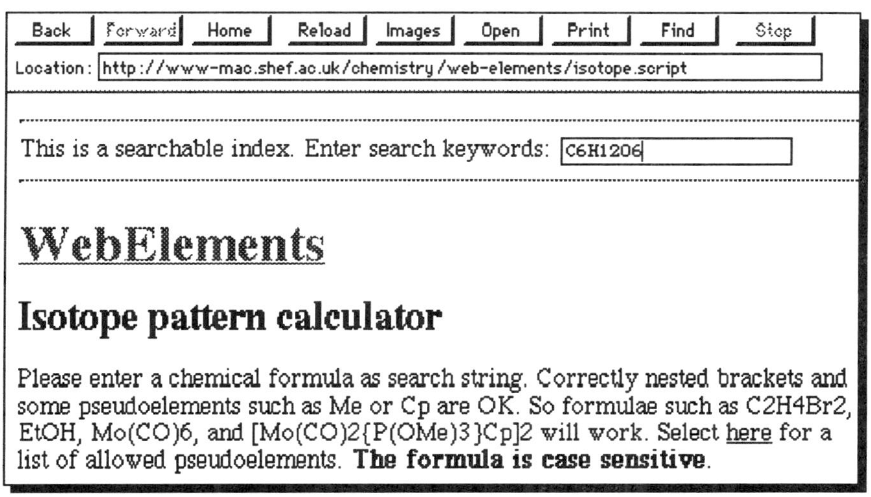

Figure 6.14. Isotopic mass ratios produced from a search query.

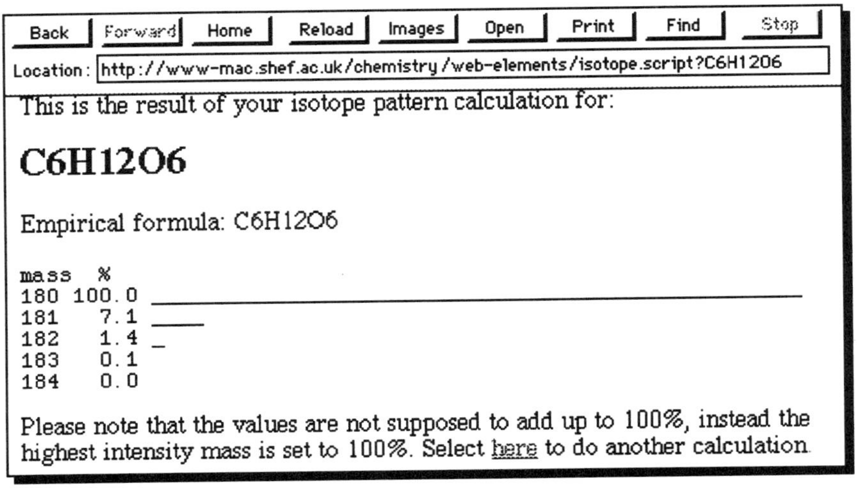

into the browser window. It is not currently possible to formulate search queries using a graphical user interface, but no doubt these will be developed in due time to enable chemical structure searches to be performed.

Local Printing, Saving, and Processing of Documents. Although the term "pages" of information is often used, this is in fact something of a misnomer, because a browser window can be of any size, and the contents are therefore essentially pageless. All browsers can nevertheless reformat data to correspond to specified page sizes and send the result along with embedded graphical images to a printer attached to the local computer. Portions of text in a browser window can also be transferred to other application windows via a copy-and-paste or drag-and-drop operation, and the *Netscape* browser also supports copy-and-paste for displayed images.

The symbolic contents of a browser window can also be saved as either a text file or as an HTML document to the local disk system. The ability to acquire a disk copy of the original HTML file also allows the reader to inspect the syntax used by the original author to achieve any particular markup effect and permits error-free access to the semantic content of the original. In the future, the semantic content of an HTML 3.0 document will also allow complex mathematical symbolism to be retrieved and converted for local use. Specifically tagged chemical information must await the development of general purpose chemistry document-type SGML definitions. Another way, however, of acquiring chemically specific information, is through the use of MIMEs (multipurpose Internet mail extensions, discussed in the next section) definitions and helper applications.

Using a WWW Browser with Helper Applications

WWW Browsers can display only text and HTML markup, and inlined images in formats such as GIF (graphics interchange format), JPEG (joint photographic experts group, with *Netscape)*, or PICT (with Macintosh-based browsers). Other file formats must be processed with external programs known as "helpers", which are in turn invoked using a mechanism known as MIME.

Multipurpose Internet Mail Extensions (MIMEs). In the preceding discussion, several MIME content types were implicitly introduced as information labels for a file. These labels comprise a primary and a secondary type:

```
text/html
image/gif
image/jpeg
```

`text` and `image` are two of the seven so called primary MIME content types, the others being `audio`, `video`, `application`, `message`, and `multipart`. There is also a differentiation between 8-bit binary files and 7-bit text files, which have an explicit end-of-line (EOL) character such as a `CR` (carriage return, Macintosh), `LF` (line feed, UNIX) or `CR+LF` (DOS) present. The EOL character in a text file must be interpreted by the WWW browser, which can be configured to process files in any of the seven primary MIME types as either 8-bit binary or 7-bit text types. It is a little confusing that the attribute "text" can be used both as a primary MIME type and as a general descriptor of the internal structure of a file. The `multipart` and `message` MIME types are used extensively for electronic mail enclosures, but have hitherto not been implemented for use in WWW browsers. Recently, the chemical primary MIME type has been proposed[28] to take advantage of the rich complexity and diversity of chemical information and the manner in which it is stored.

The secondary descriptor is known as a MIME media content type, and serves to define the subsequent treatment of the document. For example, `text/html` indicates the file should be treated as a document to be marked up by the browser itself, whereas `text/plain` would be treated without any markup, and `text/richtext` would be flagged as requiring the intervention of a suitable program such as Microsoft *Word* for its display. In retrieving a remote document, a WWW browser will need to identify the MIME type in order to process it correctly. This identification is done via the HTTP protocol communication between the WWW server and the browser. Thus the HTTP server will have been configured to associate a MIME type with a particular filename extension. The server configuration entry `image/gif gif` would result in, for example, `mypicture.gif` being sent from the server together with the MIME header `image/gif`.

The browser in turn must be configured to associate the header `image/gif` with the program to be used to visualize the image. Browsers are expected to process `image/gif` or `text/html` media content types themselves, but types such as `audio/basic`, `video/mpeg`, `application/compressed`, or `chemical/x-pdb` are not currently supported by WWW browsers and hence must be associated with external helper programs, presumed resident on the user's computer. For each type, a default action, such as saving the file on the local hard disk, should also be defined.

Helper Programs. These are programs that are resident on the local computer system and can be invoked from a WWW browser to perform specific tasks such as displaying a graphic image, playing a sound or animation file, unpacking a program archive, or processing a molecular datafile.[29] Acquiring and implementing these programs is the responsibility of the user. Some of the freely distributed programs are indicated in Table 6.3.

The configuration of the WWW browser to recognize these programs is also highly platform-dependent. This configuring can be done through a specific menu item as shown in Figure 6.15, and the selections are then written to the browser preferences file.

On UNIX systems, a more general definition known as `mailcap` is used to define the relationship between browsers and helpers. A typical set of entries such as

```
image/gif; xv %s
chemical/x-pdb; rasmol %s
```

Table 6.3. Helper Programs for WWW Browsers

MIME Association	*X-Windows*	*MS Windows*	*Macintosh*
application/ compressed	*tar*	*winzip*	*Stuffit Expander*
image/...	*xv__*	*Lview Pro*	*JPEGView, Graphic Converter*
audio/basic	*playaudio*	*WPLANY*	*SoundMachine*
video/mpeg	*mpeg_play*	*mpeg_player*	*Sparkle*
application/pdf	*Acrobat Viewer*	*Acrobat Viewer*	*Acrobat Viewer*
text/postscript	*gs*	*GhostScript*	*Mac GhostScript*
chemical/...	*RasMol, XMol, Molfile Viewer,. EyeChem,* etc	*RasWin, Molfile Viewer, MAGE,* etc	*RasMac, Molfile Viewer, MAGE,* etc.

Figure 6.15. A typical configuration page for helper programs.

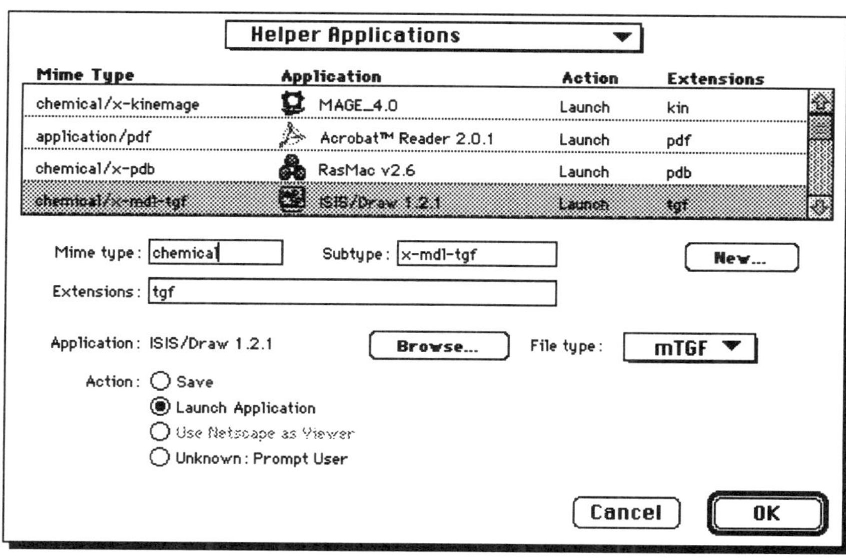

is contained in a file called `.mailcap` in the user's home directory or is allowed to default to a system-wide definition stored in a location such as

```
/usr/local/lib/mosaic/mailcap
```

Although the `mailcap` formalism[30] allows for a complex sequence of actions to be performed, in most cases the simple definitions shown suffice. The syntax simply means that when a UNIX-based browser receives a command from an HTTP server indicating that a particular hyperlink corresponds to, for example, the `chemical/x-pdb` MIME type, a local program called *RasMol* will be invoked, and the %s will be substituted for the name of the remote file the browser is retrieving. On MS Windows systems, files with the suffix `.ini` are used to define the MIME types, such as `Mosaic.ini`

```
TYPE11="chemical/x-pdb"
chemical/x-pdb=c:\raswin\rasmenu.exe %ls
chemical/x-pdb=.pdb
```

Suffix mapping is also defined on the browser to enable the recognition of local disk-based files and is not needed for HTTP-based server transactions for which the MIME-type information is provided by the server alone. A test page for checking that your local configuration is correct is available[31] (Figure 6.16).

Normally, each specific application program invoked from a WWW browser by a MIME association will display information in a separate window on the screen. Other than the undesirable aspect of having too many screen windows open, there is another unfortunate consequence of the way helper programs are currently implemented. For example, the helper program is unlikely to support the HTTP or HTML protocols, and so cannot itself participate in any further hyperlink transactions. In essence, the data on display is in an information cul-de-sac. This would not matter for image types, which have no defined mechanism for embedding hyperlinks, but it would be restrictive for chemical or VRML data types, for which hyperlinks in the form of URL or URN entries can easily be embedded into the data content. Newer paradigms based on document- rather than application-centric models promise a solution for the future. For example, in the OpenDoc formalism, a compound document is assembled from small component "parts", and the interdocument communication is handled transparently by the operating system. OpenDoc parts providing HTTP, HTML, and VRML support combined with molecular

Figure 6.16. A test page for checking MIME configurations.

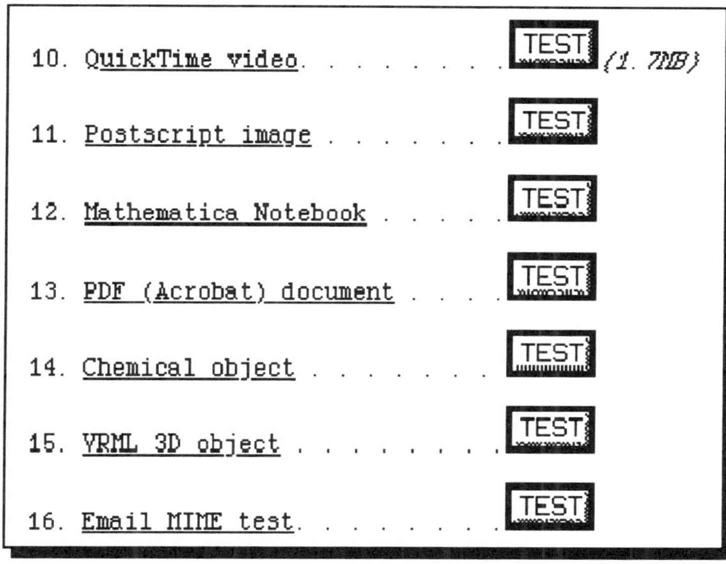

visualization and analysis parts would remove the need for the user to acquire a plethora of separate "monolithic" programs.

World Wide Web Servers

Whereas WWW browsing is essentially a personal activity with little implication for others, setting up a WWW server carries more responsibility and requires greater attention to copyright and other legal matters. Although it is relatively easy to set up servers on all three of the main computing platforms, the configuration in each case is significantly different. In what follows therefore, only a brief introduction to the technical aspects of UNIX-based servers is given. and we concentrate on the aspects of preparing chemically oriented material for HTTP servers.

Acquiring and Configuring a WWW HTTP Server

The majority of WWW servers[32] are implemented on UNIX systems, if only because these systems tend to remain on 24 hours a day, are systematically backed up, and are the best developed technically. Freely available installation packs containing the executable server code are available from NCSA, CERN, and their mirror sites for a variety of UNIX software (Table 6.4), and a Webmaster starter kit is available via the URL **http://www.eit.com/wsk/doc/** with a Forms interface for customization. In general an installation consists of acquiring a save set in `tar` format using FTP-type file transfer, unpacking it into a suitable directory, and customizing a number of configuration files for the local site. The Webmaster starter kit automates much of this procedure. Customization is via a suitable configuration file that establishes the server name and various directory paths for the HTML and/or VRML documents it is going to serve. To start the Web server, the HyperText transport protocol daemon (a program process that runs continuously and can spawn copies of itself) must be run, using a command of the type

```
/usr/local/etc/httpd/httpd
```

The system should be configured to ensure this daemon is started correctly when the system is rebooted. Installation packs are also available

Table 6.4. Locations of HTTP Servers and Ancillary Software

URL or FTP Site	*Software*
ftp://ftp.ncsa.uiuc.edu/Web/ncsa_httpd	`NCSA Mosaic`
ftp://hackberry.chem.niu.edu/pub/Mosaic	Various
ftp://src.doc.ic.ac.uk/packages/web/	Various
ftp.ncsa.uiuc.edu/Web/ncsa_httpd/ contrib/whtp	`MS Windows`
ftp://emwac.ed.ac.uk/pub/https	`Windows NT`
http://www.w3.org	CERN
http://hopf.math.nwu.edu/	`Gopher+HTTP`
ftp://ftp.ncsa.uiuc.edu/Web/ncsa_httpd/cgi/	`CGI Binaries`
http:hoohoo.ncsa.uiuc.edu:80/cgi/	`CGI` and `Forms`
ftp://src.doc.ic.ac.uk/computing/ information-systems/wais/freeWAIS-sf	`freeWAIS-SF`
http://www.eit.com/software/wwwwais/ wwwwais.html	`WWWWAIS`
ftp://ftp.ncsa.uiuc.edu/HDF	`Hierarchical Data`
http://siva.cshl.org/wusage.html	`Wusage`

for Macintosh, MS Windows and Windows NT, and GN Gopher/HTTP, which offers both Gopher and HTTP protocols in a single package (Table 6.4). A commercial product called `Netsite` from Netscape Communications Corporation offers a number of additional features, including encrypted secure transactions and better system performance than the earlier NCSA or CERN servers.

Preparing Material for a WWW Server

Before creating a home page for a new server, some simple style guidelines are worth considering. First, decide which material will be largely for local consumption and use, and which will used as a "shopping" window to the world. A criticism of many early servers was that they were highly recursive; that is, they all pointed to each other, and only a small proportion broadcast anything of interest or, more importantly, of quality to the rest of the world. The server mission should be clearly indicated, and it should highlight material that you believe to be original and of genuine interest to others. If the WWW becomes associated with material of a low signal-to-noise content, then the entire concept will deservedly not flourish. It goes without saying that all material mounted on a Web server should be legal and should not infringe copyright or intellectual property rights. Even reproducing apparently noncontentious images such as a logo or an institutional crest can cause difficulties, because such images are often copyrighted. However, copyright law in this area is often obscure and untested.

In designing a home page, you should keep the number of in-lined images to a minimum so that the network delay in downloading is not excessive. However, a purely text-based home page does not have much of the look-and-feel element that is essential in attracting attention. A good compromise is to insert a few thumbnail images that instantly give an impression of what the home page is trying to achieve. It is also helpful to indicate the authors and last modification date of any particular WWW page; it is surprising how often this information is omitted and how frustrating it can be for others who want to contact the authors for further information. The two relatively small images in the example page in Figure 6.17 are left unexpanded and shown as small icons, with an associated text description entered via the `IMG ALT=` descriptor in the HTML language. This setup enables useful information content to be achieved for minimal network bandwidth,

Figure 6.17 An example of a Web home page.

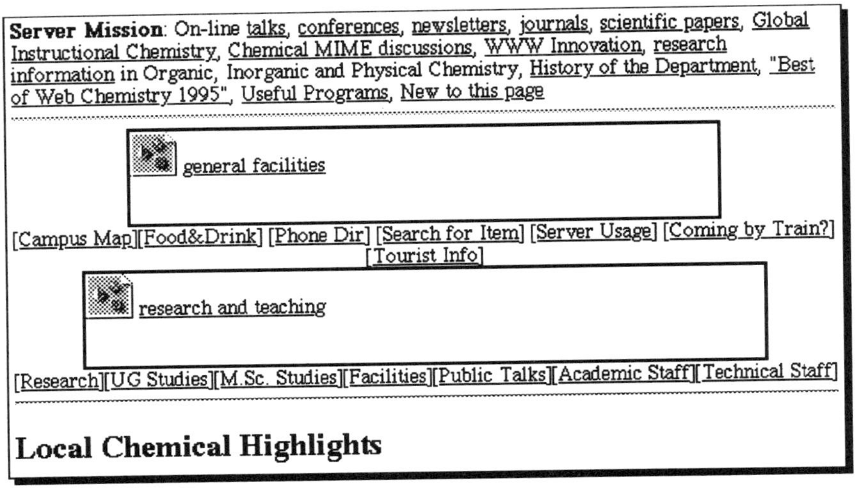

and also allows for the possibility of text-to-speech conversion to help the visually impaired.

Many individual users will also want have at least some material relating to their professional and perhaps personal interests prepared for mounting on a server. Such personal home page citations are useful for appending to e-mail signatures or to visiting cards. The first question asked by most users wanting to prepare a home page is what format the information should be provided in. A good starting point is a simple text file. Familiarity with a relatively small set of HTML formatting commands will then enable the user to create a simple list of publications with headings, italics, and bold type. Hints on appropriate style guidelines can also be obtained by seeking a well-presented existing home page on the Web, saving the content as an HTML document on the user's own local disk, and using this as a model.

The HTML Document-Type Definition. The specification of HTML is currently carried in the form of IETF (Internet engineering task force) documents. Several simplified primers are also available (Table 6.5). Because HTML has evolved rapidly during the past few years, a short historical perspective of the definition is given here to help resolve some of the early confusion (Table 6.6). The first version of HTML

Table 6.5. Programs and Documentation for HTML Preparation

URL	*Description*
http://www.w3.org/hypertext/WWW/DesignIssues/Overview.html	Designing your server
http://www.ncsa.uiuc.edu/demoweb/html-primer.html	Basic HTML primer
http://www.ncsa.uiuc.edu/General/Internet/WWW/HTMLQuickRef.html	Quick reference guide
http://www.w3.org/hypertext/WWW/Provider/Style/Introduction.html	Style guide
http://www.w3.org/hypertext/WWW/Style/	HTML 2.0 guide
http://bookweb.cwis.uci.edu:8042/Staff/StyleGuide.html	Elements of style
http://www.w3.org/hypertext/WWW/Tools/Filters.html	General tools
http://www.sq.com/hmpro.html	`HoTMetaL`, `Panorama`
http://cbl.leeds.ac.uk/nikos/tex2html/doc/latex2html/latex2html.html	`Late`χ to HTML
ftp://ftp.cray.com/src/WWWstuff/RTF/rtftohtml_overview.html	RTF to HTML
ftp://skye.aiai.ed.ac.uk/pub/tex2rtf/"	`Te`χ to RTF
http://ai.eecs.umich.edu/highc/software/translator/	Claris `Works` to HTML
http://sec-look.uiowa.edu/about/projects/arachnid-page.html	`Arachnid` formatting
http://www.ceneca.com/Products/PageMill.html	`PageMill` formatting
http://akebono.stanford.edu/yahoo/Computers/World_Wide_Web/ HTML_Converters/WPTOHTML/	`WordPerfect` to HTML

Table 6.6. A Summary of Supported Features in HTML Versions

Feature	*1.3*	*2.0*	*3.0*
Images	★	★	★
Forms	–	★	★
Tables	–	–	★
Composite figures	–	–	★
Mathematics	–	–	★
Character sets	–	–	★
Style sheets	–	–	★
Typed links	–	★	★

NOTE: An asterisk indicates that the feature is supported; a dash indicates that it is not.

that corresponded to a correctly defined SGML DTD was published in June 1993 and became known as version 1.3. This standard formed the basis for most first-generation WWW browsers, but has now largely been superseded by newer revisions.

HTML 2.0 Features. From May 1993 to May 1994 an enhanced specification known as HTML+ was defined by Raggett[33,34] in the form of two successive Internet drafts, each having an expiration period of 6 months. During the active period of this proposal, only parts of this HTML+ standard were actually implemented in any browsers by software developers. In May 1994 the decision was taken to replace the then expired HTML+ document with two separate definitions, to be known as HTML 2.0 and 3.0.

HTML version 2.0 was published in November 1994 and represented essentially a consolidation of those components of HTML+ implemented in Version 2.4 of the NCSA X-Windows `Mosaic` browser. Conspicuous by their absence were several features considered desirable for chemistry, such as Greek and mathematical characters, superscripts and subscripts, and support for tables and equations. Some of the special characters are actually available in the so-called ISO Latin 1 character set, but a general solution to implementing other characters sets was expected to be addressed in a version 3.0 of the HTML standard, expected in 1995.

In addition to the published definition of HTML 2.0, browsers such as `Netscape` introduced a number of proposed extensions such the <CENTER> tag. Because of the well structured nature of the SGML–HTML language, such tags should simply be ignored by any browser that does not support them. Originally criticized for introducing nonstandard extensions, the approach followed by `Netscape` may be a forerunner to how subject-specific browsers might implement rich subclasses of markup that might be too expensive to install in more general purpose browsers.

Expected HTML 3.0 Features. The principal new components of the next standard specification, which was expected to emerge during 1995, are support for table formatting and much greater control of text flow, alignment, and size attributes of both text and images. The `Arena` browser implements much of this as a "test-bed" for the new standard. Mathematical setting is also implemented in this browser, and further extensions such as style sheet implementations and client-side ISMAPs were expected to be added. HTML 3.0 is also expected to contain additional tags to make documents more accessible for information-searching programs.

The Basic Structure of an HTML Document

Although in practice most browsers can make a reasonable attempt to display sensibly even a totally untagged text file, adherence to recommended guidelines is highly desirable. A fairly inclusive example of the general structure, corresponding to the well-defined HTML 2.0 standard available in 1994, is as follows. In future revisions of HTML, further elements and attributes may well be added.

```
<!DOCTYPE HTML PUBLIC "-//W30//DTD W3 HTML 2.0//EN">
<HTML>
<HEAD>
<TITLE> Document Title </TITLE>
<BASE HREF="http://your.machine/"></BASE>
<LINK HREF="http://your.machine/other.document">
<META EXPIRES HTTP-EQUIV="Expires"> Tue, 04 Dec 1995
21:29:02 GMT </EXPIRES>
<META HTTP-EQUIV="Keywords" CONTENT="Chemistry, Organic">
</HEAD>
<BODY>
<!— This is a comment line followed by the body of the
document —>
<!— This is a comment line preceding a horizontal rule —>
```

```
<HR>
<ADDRESS>Your address</ADDRESS>
</BODY>
</HTML>
```

!DOCTYPE: This header identifies the precise SGML document-type definition (DTD) used in conjunction with an SGML parser, in this case version 2.0 of the HTML definition.

HTML: In principle, a document could contain several different document-type definitions, differentiated by this element.

HEAD: The header section of a document can be parsed separately from the body, in recognition that some preliminary negotiation between an HTTP server and an HTML browser serves to establish attributes such as the size of a document and perhaps whether a particular style for the document might need to be downloaded.

TITLE: This is used for the window title by most browsers, but also fulfills a more important role in defining one of the search terms for the various Internet-wide indexing agents, which can be expected to play an increasingly vital role in bringing structure to WWW information.

BASE: Many URL addresses embedded in HTML documents employ a relative naming scheme, in which a document is assumed to reside on the same server as the "parent" that referenced it. If the parent document is saved by a user on his or her own machine, all the relative URLs would no longer be correct. The base reference ensures that such relative URLs are no longer orphaned by associating them with the original source of the document.

LINK: This indicates a relationship between the document and some other object such as authorship, related indexes, and older or more recent versions.

META: This is used to embed document information that is not defined by other HTML elements.

HTTP-EQUIV enables the header to be processed according to well-defined syntactic mapping. In the example shown, the expiration date of the document or defined keywords can be extracted from the document without having to download the entire document from the server to the browser.

BODY: The main contents of the document are inserted here.

ADDRESS: A specific tag to indicate the author of the document.

Specific examples of the many other formatting elements are not illustrated here. A good way to see how particular effects are achieved is to use a browser to acquire the source code for appropriately formatted WWW pages. In many cases, it is not necessary to manipulate the actual HTML elements if a high-level HTML editor or converter has been used. Indeed, it is estimated that a significant proportion of hand-coded HTML available on the WWW system during 1995 contained significant semantic errors. The relatively simple and forgiving nature of the first generation of WWW browsers meant that these errors were often not revealed. In future, as the HTML gets more complex, it will become much more essential to ensure the correctness of the documents produced.

HTML and SGML Converters. A number of suitable converters have been produced in recognition that many putative HTML documents probably already exist in other marked-up formats such as Microsoft *Word*, Claris *Works*, *WordPerfect*, *FrameMaker*, or *LaTeχ* (Table 6.5). The HTML produced by these conversion programs currently mostly corresponds to the HTML 1.3 standard rather than to later versions. One more interesting development lies in the conversion of documents encoded using the ISO 12083[4] SGML document-type definition for scholarly publishing, a standard being adopted by many publishers for the storage, archiving, and indexing of scientific publications. Currently this standard is semantically much richer than HTML 2.0, and any conversion process is therefore bound to result in the loss of information. In one implementation,[35] the translation of mathematical markup is done by creating in real time a GIF image of the equations using TrueType fonts, and producing HTML code with hyperlinks to the images for display on the browser. The HTML instructions and the GIF images are never explicitly stored anywhere, but are generated as required. An alternative approach is not to convert the SGML-encoded documents at all, but to view them locally using an SGML helper program, and to reserve HTML for more basic markup and HyperText embedding.

HTML Editors. Several commercial programs (e.g., *HotMetaL*, *WebMagic*, *CyberLeaf*, and *AnchorPage*) in effect are SGML editors with built-in support for the HTML DTD. These programs allow the syntax of an HTML file to be checked by the editor, and new entries to be made that are semantically correct. Use of such an editor in effect requires no knowledge of HTML and produces a "what-you-see-is-what-you-get" effect on the screen. This does mean that more advanced features of HTML 3.0 would have to await the production of the corresponding HTML 3.0 DTD to be implemented into such programs. The second approach is to use a "plug-in" in conjunction with a word processor such as Microsoft *Word*, Claris *Works*, or *WordPerfect* or publishing programs such as *Pagemaker* or *Quark Express*; a plug-in adds specific tools to enable hyperlinks, styles, or in-lined images to be added to a document and allows the result to be saved as a series of HTML and graphical files. Using this approach, it should not be possible to create incorrect HTML commands. This is perhaps the most transparent approach to authoring new documents, and is likely to be extensively supported by commercial developers. For example, a custom program called PageMill[36] will allow elaborate design of HTML documents in a "what-you-see-is-what-you-get" sense.

The third type of editor corresponds to a lightweight text-handling program in which the raw HTML is visualized, but using HTML specific extensions to insert appropriate tags and to perform preview functions. An example of this is the extensions available for the Macintosh *BBedit* program (Figure 6.18).

Figure 6.18. HTML extensions implemented in a text editor.

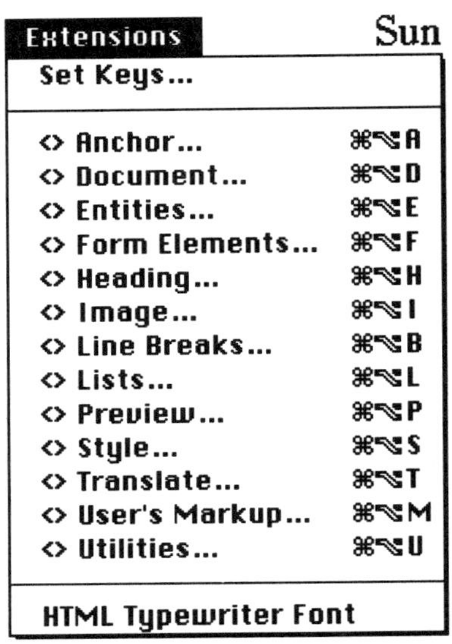

Such extensions also allow a preview of the formatted HTML and a check of the syntax. The advantages of using such programs are their speed and the fact that new extensions to HTML can be introduced very quickly.

Hyperlinking a Document

Sections of text that define the beginning and end of a hyperlink are called "anchors" and are defined by the markup

```
<A options> ...text... </A>.
```

This type of construction is also sometimes referred to as a "container", and requires in most instances that the container be enclosed with appropriate start and end elements. Hyperlinks can easily be inserted transparently into an HTML document by using appropriate editors, but to illustrate the concept, six explicit examples of the syntax are given here.

1. ```
   <A HREF="gopher://pdb.pdb.bnl.gov/11/">
   Brookhaven Protein Data Bank</A>
   ```
   Here, one can see the general format of an HTML tag, whereby `&lt;action&gt;` and `&lt;/action&gt;` enclose a section of text to be marked up. In this case, the text will be made a hyperlink, which, when selected using a browser, will invoke the URL pointing to a gopher server.

2. ```
   <IMG ALIGN=middle ALT="The Boss"
   SRC="http://www.ch.ic.ac.uk/photo.gif" >
   ```
   This command inserts an in-lined image file acquired from an HTTP server. The `ALT` field can be used as an alternative to displaying images where the network connection is too slow.

3. ```
   <A HREF="http://www.ch.ic.ac.uk/atp.pdb">
   <IMG ALIGN=BOTTOM ALT="PDB Format"
   SRC="http://www.ch.ic.ac.uk/atp.gif"></A>
   Hyperactive Molecules
   ```
   This syntax illustrates how an in-lined image can itself be made a hyperlink. The .pdb filename extension implies the use of a chemical/x-pdb MIME type that the user must associate with an appropriate helper program. When the image itself is selected, the connection to the HTTP server will result in pdb format coordinates
   ```

being transferred across and transformed into a three-dimensional diagram by the helper program.

4. ```
<A HREF="http://www.ch.ic.ac.uk/cgi-bin/imagemap/
leachax">
<IMG ALT="clickable map"
SRC="http://www.ch.ic.ac.uk/diol.gif"ISMAP></A>
```
Here, the hyperlinking is extended to the image itself by adding an ISMAP label. When the cursor is placed upon the image, the pixel coordinates of the cursor (relative to the top left of the image) are transferred to a coordinate mapping program called *imagemap* located in the cgi-bin directory of the HTTP server. This program relates these coordinates to a set of "hot-spots" in the image defined in a map file stored on the HTTP server, in this instance called leachax.map. The location of this map file has to be defined in the imagemap configuration file imagemap.conf. The map file has the appearance

```
default http://www.ch.ic.ac.uk/.index.html
rect http://www.ch.ic.ac.uk/diol.mop 154,474 202,502
```

indicating in this case that a certain rectangular area corresponds to a specific URL hyperlink. The default entry indicates the action to be performed if no hot spot is selected. The effect achieved in this particular example is to associate specific regions of an isometric projection of a potential energy diagram (diol.gif) with sets of molecular coordinate files in pdb format. The use of chemical MIME will allow these coordinates to be downloaded when the appropriate region of the energy diagram is clicked, and visualized using the appropriate program. A future version of the HTML 3.0 standard is expected to allow such mapping to be performed entirely within a browser by replacing the ISMAP image attribute with a local function. This system would be particularly appropriate for extensive graphical material held on a local storage device such a CD-ROM or a hard disk. The proposed syntax for such client-side mapping introduces a new tag called <FIG>, which substantially improves upon the original <IMG> tag described earlier in this chapter:

```
<FIG SRC="reactionscheme.gif">
<H1>Synthesis of Taxol</H1>
    <UL>
    <LI><A HREF="step1.html" SHAPE="rect
30,200,60,16">First step</A>
    <LI><A HREF="step2.html" SHAPE="rect
```
```

```
100,200,50,16">Second step</A>
    </UL>
</FIG>
```

5. `<A HREF="telnet://stn.fiz-karlsruhe.de"> Click here for CAS </A>`
   One specific type of hyperlink is used to invoke a terminal-based service. A separate Telnet program must be available to the user.

6. `<A HREF="mailto:dummy@no.such.host"> Send a message </A>`
   The final type of hyperlink allows an e-mail message to be composed and sent. This presupposes that an e-mail handler conforming to the SMTP standard is available to the user. In general it is not possible to read e-mail messages by the same mechanism.

Maintenance of Hyperlinks

One particularly important feature of preparing an extensively hyperlinked document is in testing and maintaining the functionality of the links. The HTML specification does not readily allow the user any way of ensuring that all links in a particular document have actually been visited, a particular problem with intricate or legal documents, for which such assurance can be important. One tool in the Webmaster starter kit[37] traverses a document by automatically following all the links and reporting any errors, but it is only currently available on UNIX systems. Another commercially available tool that is designed for complete maintenance of Web documents is *`SiteMill`*, which will even automatically rename all the pointers in files that rely on any altered documents.[36]

Creating and Mapping a Clickable Image

Although more than 50 image formats are known, the common standard for WWW use is the GIF format in either the `GIF87a` or `GIF89` specifications. The more compact JPEG format will also be widely supported in the future, and an image standard customized for good network characteristics, called PNG (portable network graphics) is likely to become particularly important for chemistry. PNG can contain embedded chemical information such as atom coordinates, connection tables, and even URLs to other information sources.[38] A wide variety of both shareware and commercial programs are available to generate and process such images, although the following three suggestions normally prove adequate for most purposes: `xv` (X-Windows), `Lview Pro` (MS Windows), and `GraphicConverter` (Macintosh). Each provides the ability to read graphical files in most of the common formats, to edit them at the pixel level if necessary, to change attributes such as the color depth, and to save them in the 8-bit GIF or the JPEG standards. Two particular attributes that have become commonly used recently are "transparency", which allows one of the 256 color indices of an 8-bit GIF image to assume the background color used by the browser, and an "interlaced" format, which allows the perception of an image to be built up more rapidly by initially displaying only alternating horizontal lines of an image. These features, and others, are also expected in the PNG format, as already noted.

Some major sources of graphical images are

- from 35-mm film, using the Kodak Photo-CD process, or the new generation of digital cameras.
- digitally generated images from suitable application programs. Most molecular modeling applications are capable of saving diagrams in a suitable format. Collections of clip art also can be acquired if necessary.
- a snapshot or screen capture of an area of a window.
- a flatbed scanner.

When selecting an image for use on a WWW server, do bear in mind that the larger images (e.g., >100 kbytes) may take an inconveniently long time to download across a congested network. In this regard, do not save an image with 24-bit color depth if 8-bit or less will do, because the size of the file is directly proportional to the bit depth. If you do need greater color depth, you will have to use the JPEG specification, rather than GIF. A compromise that is often adopted is to create a small (32- by 32- or 64- by 64-pixel) 8-bit thumbnail image that will give the user an idea of the image, and then to hyperlink this image to a larger version with perhaps more color depth, which the user can select to download. Even a photograph can be recognizable as a 32- by 32-pixel image! Attention to the resolution of the original is also needed. For example, a scanned image obtained at 300 dpi or higher can only ever be displayed at 72–96 dpi on most computer monitors. Thus a 300-dpi image will be displayed magnified ≈×4 unless the resolution of an image is reduced using a suitable manipulation program.

If you wish to prepare hyperlinks within an image, you will have to use a program capable of giving the pixel coordinates of any region of the image. By convention, the top left of the image is 0,0. It is possible, although rather tedious, to map rectangular or circular regions of the image by hand. Also, programs such as MapEdit and WebMap can read an image file in PICT or GIF format and prepare a so called `.map` file suitable for reference within the ISMAP tag. The final step is normally to configure the `IMAGEMAP` program located in the `cgi-bin` of the HTTP server with the location of the `.map` file. In the future, the image mapping directives most likely will be stored instead within the HTML document and resolved by the client browser rather than the server as at present.

Audio or Sound

Some browsers are now offering the ability to record or play audio files without the need for an external program. Computers vary widely in their support for audio, and as with images, a wide variety of formats for storing the sounds are in use. Experience suggests that the so-called `SUN ulaw` format is the most suitable for cross-platform use. A number of free programs will interface between a microphone attached to the computer to allow the recording to be saved in this format. On a Macintosh computer, SoundMachine is suitable. As with images, have a regard for the remote user with a slow network connection. It is also becoming more common for music files in MIDI format to be included in hyperlinks, although whether such application is useful to chemistry remains to be established.

Animation

The accepted standard here is MPEG (Moving Picture Experts Group) format, although Apple *`Quicktime`* is certainly the more technically advanced (in 1995) in offering synchronized sound with the audio images, efficient playback without special hardware, and a virtual reality option for generating navigable three-dimensional representations. The advantage of MPEG is that smaller file sizes than the equivalent *`Quicktime`* format can be achieved. Animation offers particular advantages in the area of chemistry, as, for example,

- imparting a three-dimensional perspective to a molecule by rotating it slowly around an axis
- revealing all the features of a complex three-dimensional property, such as a wave function or electrostatic potential

- allowing the effective display of multidimensional data by using the time component as an effective fourth dimension, or indeed display of original data for which time dependence was an intrinsic component

A video animation file can be created in several ways. The simplest method involves creating a "cartoon" comprising ≈20–150 separate images taken in the same reference frame, and from which a single animation file can be created and compressed using special programs (such as `Sparkle` on a Macintosh computer). Many molecular modeling programs can also generate and save such a sequence internally. Furthermore, some background applications automatically take a specified number of snapshots of an area of the screen every second, and eventually compress them all into an animation file. This application is useful when showing how a particular program is used in the form of a tutorial. The other principal method is to acquire images from a camera, and via a digital video capture board, to generate a movie file.

Animation is one of the more innovative applications of the Web, because they provide information in a form quite impossible using the more conventional printed page. The price to be paid, however, is the network transfer times incurred as a result of the sizes of the files involved. These range from about 200 kbytes for approximately 50 frames showing a molecule revolving about an axis, to more than 4 Mbytes for 10 seconds of "talking heads". Given the state of the Internet in 1995, the transfer of any animation of >1 Mbyte in size should be considered distinctly unfriendly!

Tables, Symbols, Superscripts, Subscripts, and Mathematical Setting

Of particular interest to chemists are formatting features such as tables, equations, and Greek and mathematical characters, superscripts, and subscripts. In version 2.0 of HTML and earlier, these are not supported, and work-arounds have to be adopted. These normally amount to taking a screen snapshot from within a window of a program that is capable of displaying these attributes and then inserting a suitable graphical image into the required document. The disadvantage of this method is that the images do not rescale as the size of the browser window is adjusted or when the display font size is altered. With the introduction of HTML 3.0 and beyond, and of browsers gradually supporting this standard, these attributes can be marked up directly. Several browsers currently support both superscripts and subscripts using the tags

```
CH<SUB>3</SUB><SUP>+</SUP>
```

Some special symbols and other characters can be currently specified in terms of the ISO Latin names[39] enclosed with two reserved combinations, the & and the ; characters. Such a construction is called an entity. Thus markup tag delimiters themselves such as the > character are specified as `>` or by using the decimal equivalent `>` (being the form used by the *Netscape* browser). Other chemically useful entities found in the Latin character set are shown in Table 6.7. In general however, such characters are not always independent of the font used by the browser, and the general solution to the character-set problem expected to be implemented during 1995 may differ significantly from the detail shown here. In the future, we might even expect chemically specific HTML entities to be introduced as part of an SGML DTD definition for chemistry markup. One example of such a DTD, called chemical markup language, has recently been introduced.[40]

A table can be formatted using the following HTML 3.0 markup commands. The tags <tr>, <th>, and <td> stand for table row, table header, and table data, respectively. The closing tags (e.g., </tr>) are not mandatory, but they help to better identify the various fields visually.

Table 6.7. Character Entities in *Netscape* 1.1N or *Mosaic* 2.00b12

‚	ƒ	„	…	†	‡	ˆ	‰	Š	‹	Œ		Ž
≠	∞	≥	Σ	∫	Ω	√	≈	…	—	‘	Ÿ	⁄
		‘	’	“	”	•	–	—	˜	™	š	›
∂	Δ	Œ	‚	„	‰		ˆ	˜	˘	˙	˚	˝
ž	Ÿ	¡	¢	£	¤	¥	¦	§	¨	©	ª	«
ı	ƒ	¡	¢	£	¤	¥	œ	§	¨	©	!	«
¬		®	¯	°	±	²	³	´	µ	¶	·	¸
¬	–	®	¯	°	±	”	“	´	µ	¶	·	¸
¹	º	»	¼	½	¾	¿	À	Á	Â	Ã	Ä	Å
’	º	»	π	Π	≤	¿	À	Á	Â	Ã	Ä	Å

```
<table border>
<caption><b>Table 2. Calculated Isotope Effect</b></caption>
<tr><th rowspan=2>System</th><th colspan=2>Method</th>
<th rowspan=2>Literature<br>reference</th></tr>
<tr><th>Semi-empirical </th><th><I>Ab initio</I></th></tr>
<tr><td align=left><A HREF="struct1.cd"><IMG
SRC="struct1.gif"></A></td> <td>1.9</td> <td>3.1</td>
<td>A.Body,<I> J. Am. Chem. Soc.</I>, 1996, 222.</td></tr>
<tr><td align=left><IMG SRC="struct2.gif"></td> <td>5.8</td>
<td>6.2</td> <A HREF="http://machine/paper.html">N.O. Body,
<I>J. Perkin Trans.</I>, 1998, 222</a></td></tr>
</table>
```

The hyperlink is present in the tabular data itself, and the a graphical image is included as part of the table. The result is shown in Figure 6.19.

A proposed enhancement to be included in the HTML 3.0 specification is mathematical markup. The following HTML

```
<math>
y=sum_i=1_^n^{1<over>1+x_i_}
</math>
```

would produce the result shown in Figure 6.20.

Mathematical markup can also be included in tables, and thus quite complex scientific constructs can be attempted. Programs that cope with the interconversion of such symbolic notation embedded in SGML documents, Mathematica Notebooks, Teχ, and HTML are likely to evolve as the standards become accepted.

Figure 6.19. A table layout in *Netscape*.

Table 2. Calculated Isotope Effect

System	Method		Literature reference
	Semi-empirical	*Ab initio*	
	1.9	3.1	A.Body, *J. Am. Chem. Soc.*, 1996, 222.
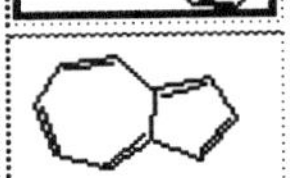	5.8	6.2	N.O. Body, *J. Perkin Trans.*, 1998, 222

Figure 6.20. Mathematical layout in *Arena*.

$$y = \sum_{i=1}^{n} \frac{1}{1 + x_i}$$

User Input via Forms

Forms[41] provide a method of sending information from the browser to a server to achieve a specified action. An example of the syntax follows:

```
<form method="POST" action=
"http://www.ch.ic.ac.uk/cgi-bin/service_email">
<INPUT type=hidden name="Request Number" value="services">
<ol>
<li> Please select service of interest : <select
name="service">
<option selected> 1000 MHz NMR<option> Five Minute single
Crystal X-Ray
</select></li><HR>
<li> Indicate the nature of your query <textarea
name="Comment" Rows=2 cols=60 >Please run a spectrum
</textarea></li>
<li> Please enter your name and e-mail address : <input
name="Name" type="text" size=40> </li>
<li><input type="submit" value="Click to Send request"><input
type="reset" value=" Click to reset "></li>
</ol></form>
```

The variables gathered from this form, which are in this example service, comment, and name, are passed to a program or script called `service_email`, which must be written specifically to process these variables, in this instance to create an e-mail message to be sent to the manager of the selected service. It is to be hoped that the ad hoc nature of programs such as `service_email` will in the future be replaced by more general mechanisms for processing Forms. Nevertheless, the scope of the Forms mechanism is wide. For example, it becomes possible to create background batch jobs to perform a quantum mechanical calculation or a database search. If the output from such a task is to be returned to the Web browser, it will need to be marked up in HTML beforehand, if necessary by another appropriate program.

A more sophisticated example of Forms in action shows how a search query can be invoked and the results formatted using a program called *wwwwais* with the results returned to the user. The following HTML code produces the screen display shown in Figure 6.21.

```
<FORM METHOD=GET ACTION="http://your.machine/cgi-bin/wwwwais">
Search for:
<INPUT NAME="isindex" VALUE=""><INPUT TYPE="submit"
VALUE="Search">
<INPUT TYPE="reset" VALUE="Clear selections"><br>
Sort by:<SELECT
NAME="sort"><OPTION>score<OPTION>lines<OPTION>bytes
<OPTION>title<OPTION>type</SELECT>
Max hits:
<SELECT NAME="MAXHITS"><OPTION>20<OPTION>100</SELECT>
</FORM>
```

This, in turn, produces the result shown in Figure 6.22.

Defining Server MIME Types

MIME types are normally defined in a file such as `mime.types` on the NCSA HTTP server or `httpd.conf` on the CERN HTTP server, and for most purposes need not be altered. If, however, you wish to add new MIME types appropriate for chemical visualization, you will have to add entries of the following type for the NCSA server:

```
chemical/x-pdb pdb
chemical/x-mopac-input mop
application/x-mathematica ma
```

Figure 6.21. Forms-based entry for a search query.

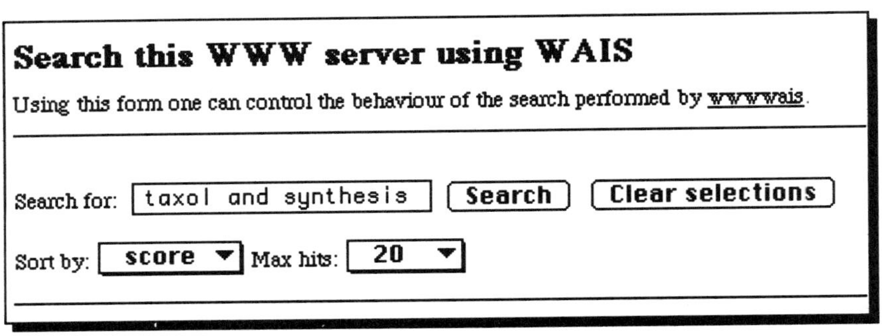

Figure 6.22. Output from a search query formatted using *wwwwais*.

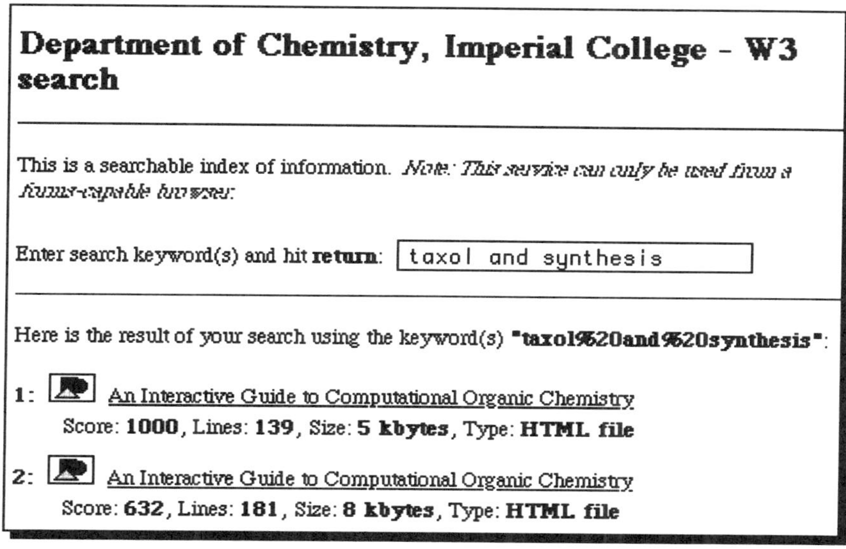

or for the CERN server:

```
AddType .pdb chemical/x-pdb 7bit 1.0
AddType .mol chemical/x-mdl-molfile 7bit 1.0
```

To indicate their provisional nature, an x may precede some of the secondary type definitions. Equivalent definitions also must be added to the browser configurations.

Defining Access Control Lists

Several levels of control allow access to a server. Each HTTP server implements this access in a unique way. For example, using the NCSA HTTP server, a file named `.htaccess` is placed in the directory in which access needs to be controlled. A typical file has the appearance

```
AuthUserFile /directory-path/.htpasswd
AuthGroupFile /directory-path/.htgroup
AuthName Journal of Chemistry Subscribers Only
AuthType Basic
```

```
<Limit GET>
order deny,allow
deny from all
allow from joc.org
</Limit>
```

which would allow access only from the specified Internet sub-network `joc.org`. Individual users can be validated with password access as follows

```
<Limit GET>
require group subscribers
</Limit>
```

where the group `subscribers` can contain any number of individual entries, each requiring a different password. Such password validation is not supported by all browsers. In the future, more sophisticated schemes based on digital signatures may be expected to evolve, and a number of commercial products such as *Netsite* support RSA encryption schemes to ensure confidential transactions.

The Server Log and *Wusage*

A log file of all access to the HTTP server is normally saved in an indigestible form not suitable for visual browsing, nor can anyone except the system administrator normally access this file. This information can be processed and presented as a formatted HTML document via the server itself using a program called *Wusage* (Figure 6.23).

Other forms of usage statistics can monitor the total number of visits to any particular server or to any individual documents held on a server, and one can even envisage a form of science citation index in which not only the quantity but the quality can be assessed.

An example of how analysis of the server log can yield valuable information is provided by the ECTOC (Electronic Conference on Trends in Organic Chemistry) conference statistics.[42] These statistics indicate the most widely read papers and show graphically a demographic distribution of the readership across continents and time zones. Readers could even request individual statistics for particular papers, together with a breakdown of what types of documents were being accessed for each paper, and even how often the access statistics themselves were requested.

Figure 6.23. Output from the *Wusage* program.

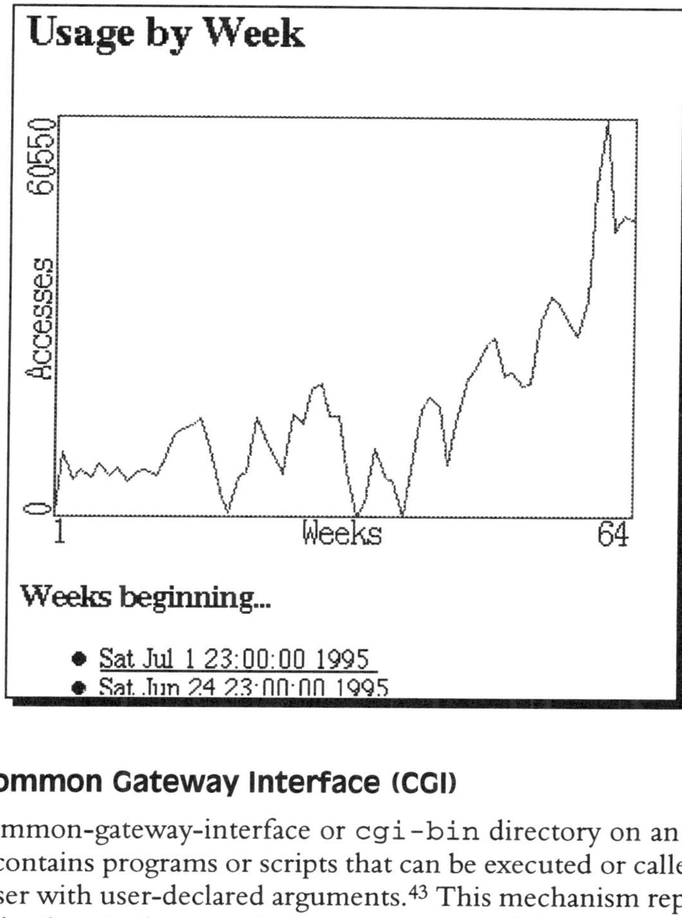

The Common Gateway Interface (CGI)

The common-gateway-interface or `cgi-bin` directory on an HTTP server contains programs or scripts that can be executed or called from a browser with user-declared arguments.[43] This mechanism represents probably the single most important aspect of the WWW system, because it is capable of generating entirely new "value added" information, rather than simply presenting previously archived material to the user. In theory at least, virtually any aspect of computational chemistry could be incorporated into this mechanism. Obvious actions could range from a key-word search on a WAIS index, the compilation of an HTML entry from a Forms request, or the execution of a molecular modeling calculation via a suitable script. In general, CGI programs are expected to respond with the delivery of an HTML document back to the browser, such that the response can appear in the browser window. An index search for example would produce a page in which all the hits are hyperlinked to the original source.

The power of the CGI mechanism also requires clear guidelines to protect the security of the user's own computer. For example, it should not be possible to directly install any sort of script or program on a client computer and have it execute transparently, or for the CGI interface to request unsuspected client-side operations such as file deletion. When mechanisms such as CGI or MIME issue a request for a client-side program or script to be executed, it is the responsibility of the user to ensure that such programs are installed in a separate operation and that the nature of such programs is trusted by the user.

Indexing a WWW Server

Text-based indexing would normally be performed in two parts. A WAIS or other indexing engine would be used to produce the inverted database structure on the HTTP server necessary for a rapid response to any indexing query. A number of free programs such as `Free-WAIS-SF`[44] and `Harvest`[45], as well as commercial programs such as `Lycos`[25], are available for this purpose. Such programs can act selectively on the header or keyword information provided for the purpose within an HTML document, resulting in rapid and often very high quality query responses. The implementation of a searchable index term into any specific document is accomplished by inserting the following into the HTML document:

```
<ISINDEX ACTION="http://your.machine/cgi-bin/search-engine">
```

The action flag indicates the URL address of the program that will be used to initiate the search and format the results for display on the screen. The search query is normally a line of simple text, which can include logical operators such as

```
taxol and synthesis
```

Improved specifications for WAIS indexing were starting to be introduced in 1994, conforming to the Z39.50 (1992) library "deep searching" specification.[46] The task of integrating Z39.50-based searches with the WWW is not entirely trivial, because WWW is based on a stateless communication protocol, whereas Z39.50 is a so-called "stateful" system in which responses generated from an earlier search query can be used in the context of future queries. The integration of HTTP and Z39.50 servers will probably be a high priority in the future. Similar growth of standards for chemically specific indexing based on SMILES strings[27] and similar specifications is also expected.

Registering a WWW Server and New Home Pages

If you want other server administrators to insert hyperlinks in their WWW pages to documents on your own server, it is necessary to notify various key sites that you are in operation. There are no absolute rules on how this is done, but the following hints may help.

- Each country normally has one site keeping an index of all known servers in that country. An e-mail message to the coordinator with a short description may suffice.
- Chemistry as a subject has several global home pages[47], whose administrators should be contacted if you wish your entry to be included.
- A number of systems regularly index the globe according to certain criteria. A meta-index of all such search engines allows you to locate other chemistry servers by a keyword search. Connect to the URL **http://cui_www.unige.ch/meta-index.html** for further details.
- The NCSA "What's New" page[48] is generally regarded as the best forum for announcing a new server or an interesting page. This has a Forms-based interface for registering new items.

Encryption and Authentication

In many instances, secure transactions[49] between a user and some remote service are required, as for example in the use of credit cards for conference registration or perhaps the transfer of more sensitive documents. An enhanced version of the HTTP protocol, to be known as Secure-HTTP, or more simply SHTTP[50] will allow a variety of encryption and other security schemes to be implemented. In parallel, extensions to the basic HTML markup that have been proposed would allow the user to control the nature of the security embedded within a document. Thus a URL of the type

```
shttp://www.anywhere/most_private.html
```

is likely to require the remote reader to provide at least a password to read the document. Another approach has been proposed by Netscape Communications involving a protocol known as SSL or Secure Sockets Layer.[51] Whether a single mechanism for ensuring secure transactions will emerge or whether we will have to cope with several competing protocols is not yet apparent.

Updates

Inevitably, in a rapidly evolving and maturing field in which standards are still being defined, some of the detail described here may have evolved by the time this book appears in print. A brief update home page contains information received too late for inclusion here. Enter the URL **http://www.ch.ic.ac.uk/WWW/** to access it.

Acknowledgments

I thank the many people on the Web whose comments and advice have helped to create this chapter, and Glaxo Research and Development (Greenford) for financial support.

References

[1]Kay, A. "The Man Who Made Computers Personal", *New Scientist* **1993**, June 19, p 32.

[2]Berners-Lee, T. J.; Cailliau, R.; Groff, J.-F.; Pollermann, B. "CERN, World Wide Web: The Information Universe". In *Electronic Networking: Research: Applications and Policy;* Meckler Publishing: Westport, CT, 1992, Vol. 2, pp 52–58.

[3]Berners-Lee, T. J. *HyperText Markup Language;* CERN: Geneva, Switzerland, 1993. See **http://www.w3.org/hypertext/WWW/MarkUp/HTML.html** and **http://www.ncsa.uiuc.edu/General/Internet/WWW/HTMLPrimer.html** and **http://www.w3.org/hypertext/WWW/MarkUp/MarkUp.html**

[4]"Electronic Manuscript Preparation and Markup", International Organisation for Standardisation. NISO/ANSI/ISO 12083, 1994.

[5]Berners-Lee, T. J. *WWW Names and Addresses, URIs, URLs, URNs;* CERN: Geneva, Switzerland, 1993. See **http://info.cern.ch/hypertext/WWW/Addressing/Addressing.html**

[6]For a comprehensive list, "Frequently Asked Questions about WWW", see Boutelle, T., **http://sunsite.unc.edu/boutell/faq/www_faq.html**
[7]For further details, see Berners-Lee, T J., **http://www.w3.org/hypertext/WWW/Protocols/HTTP/HTTP2.html**
[8]Bryan, M. *SGML: An Author's Guide to the Standard Generalized Markup Language;* Addison–Wesley: Reading, MA, 1988.
[9]Goldfarb, C. *The SGML Handbook;* Oxford University Press: New York.
[10]Van Herwijnen, E. *Practical SGML;* Kluwer Academic Publishers: Norwell, MA, 1990.
[11]Microsoft Corporation, **http://www.microsoft.com/msoffice/freestuf/msword/download/ia/default.htm**
[12]Novell Inc., **http://www.novell.com/wp/busapps/epub/**
[13]Weider, C.; Mealing, M., **http://www.path.net/mitra/urn.html**
[14]The WWW94 Geneva Conference proceedings are available via the URL **http://www.elsevier.nl/**
[15]The WWW94 Chicago Conference proceedings are available via the URL **http://www.ncsa.uiuc.edu/SDG/IT94/Agenda/Agenda.html**
[16]See for example the 1995 Gordon Conference on Inorganic Chemistry, **http://www.chem.ucla.edu/dept/Announce/95Gordon.html**
[17]Cheswick, W. R.; Bellovin, S. M. *Firewalls and Internet Security: Repelling the Wily Hacker;* Addison–Wesley: Reading, MA, 1994.
[18]The latest draft of the IPv6 protocol declaration is available from **http://www.ietf.cnri.reston.va.us/html.charters/ipv6mib-charter.html**
[19]Gessler, S.; Kotulla, A., presented at the WWW94 Conference, Chicago, IL, October 17–20, 1994.
[20]An overview of various proposed options for implementing style sheets is given in **http://www.w3.org/hypertext/WWW/Style/**
[21]Hermann, M., Karl Franzens University, Graz, Austria; **http://hgiicm.tu-graz.ac.at:80/958F7A8C/Chyperg**
[22]Symposia developed at INRIA (Institut National de Recherche en Informatique et en Automatique) in France; see **http://symposia.inria.fr/** for further details.
[23]Leach, C.; Rzepa, H. S., Imperial College, London, 1995. See **http://www.ch.ic.ac.uk/ectoc/**
[24]Murray-Rust, P.; Mills, A. "VSNS-PPS Course on the Principle of Protein Structure", Birkbeck College, London, 1994. See **http://www.cryst.bbk.ac.uk/PPS/index.html**
[25]Mauldin, M. L., Carnegie Mellon University, Pittsburgh, PA, Lycos, **http://fuzine.mt.cs.cmu.edu/mlm/lycos-home.html**
[26]Winter, M. J., University of Sheffield, **http://www.shef.ac.uk/uni/academic/A-C/chem/web-elements/web-elements-home.html** and **http://www-mac.shef.ac.uk/chemistry/web-elements/**

isotope.script and **http://www-mac.shef.ac.uk/chemistry/web-elements/percent.script**

[27]Weininger, D. "The SMILES Format", *J. Chem. Inf. Comput. Sci.* **1988,** *28,* 31. See also **http://www.daylight.com/kisstxt/tutorials/ smiles/ smiles.html**

[28]Rzepa, H. S.; Murray-Rust, V; Whitaker, B. J. "Internet Draft: Chemical Mime Type", March–August 1995, **http://www.ch.ic.ac.uk/ chemime2.html** This proposal had not been ratified as of August 1995, and the final form adopted as a standard may not correspond precisely to that shown here.

[29]Rzepa, H. S., **http://www.ch.ic.ac.uk/chemical_mime_config.html** Associated helper programs can be acquired via **http://www.ch.ic.ac.uk/ectoc/**

[30]Borenstein, N.; Freed, N. Internet Request for Comment (RFC) No. 1590, 1994. See also **http://www.ncsa.uiuc.edu/SDG/Software/ Mosaic/Docs/mailcap.html**

[31]Nelson, S. D. "E-mail MIME Test Page", Lawrence Livermore National Laboratory, Livermore, CA, 1994. See **http://www-dsed.llnl.gov /documents/WWWtest.html** and **http://www-dsed.llnl.gov// documents/tests/email.html**

[32]Friesenhahn, R. *Byte* **1995,** April, p 83.

[33]Raggett, D.; Hallam-Baker, P.; Connolly, D., **http://info.cern.ch/hyper-text/WWW/WorkingNotes/WWW94/html_plus_ws.html**

[34]Berners-Lee, T.; Connolly, D., **http://www.hal.com/~connolly/html-spec/** A guided tour around the proposed HTML 3.0 specification by D. Raggett is found at **http://info.cern.ch/hypertext/WWW/Arena/ tour/contents.html**

[35]Weibel, S.; Miller, E.; Godby, J.; LeVan, R. Abstracts of WWW Conference in Chicago, IL, October 17–20, 1994.

[36]For example, **http://www.ceneca.com/**

[37]Weber, J. C., Enterprise Integration Technologies Corporation, **http://www.eit.com/**

[38]Boutelle, T., **http://www.w3.org/hypertext/WWW/Graphics/PNG/ Overview.html**

[39]Pesce, M.; Bell, G.; Parisi, A. "VRML 1.0." See also **http://www.wired. com/vrml** and **http://www.eit.com/vrml** and **http://www.sd.tgs.com/ ~template/** For a collection of molecular entities encoded in VRML, see Rzepa, H. S.; Casher, O., **http://www.ch.ic.ac.uk/VRML/**

[40]Murray-Rust, P.; Rzepa, H. S.; Leach, C., Abstract 40. Presented as a poster at the 210th ACS National Meeting, Chicago, IL, August 1995. See **http://www.ch.ic.ac.uk/cml/**

[41]Grobe, M., University of Kansas, 1993, **http://kuhttp.cc.ukans.edu/info/ forms/forms-intro.html**

[42]*Electronic Conference on Trends in Organic Chemistry (ECTOC-1) CD-ROM;* Rzepa, H. S.; Goodman, J. M., Eds.; Royal Society of Chemistry Publications: London, 1995. See also **http://www.ch.ic.ac.uk/ectoc/accesses.html**

[43]McCool, R., National Center for Supercomputing Applications, 1994, **http://hoohoo.ncsa.uiuc.edu/cgi/**

[44]Pfeifer, U.; Huynh, T., University of Dortmund, 1994, **http://ls6-www.informatik.uni-dortmund.de/freeWAIS-sf/README-sf.html**

[45]Schwartz, M., University of Colorado, **http://harvest.cs.colorado.edu/brokers/www-home-pages/query.html**

[46]For further information on Z39.50, see **http://vinca.cnidr.org/protocols/z3950/z3950.html**

[47]Kovpelevich, M., Department of Chemistry, University of California, Los Angeles, CA, **http://www.chem.ucla.edu/chempointers.html**

[48]O'Reilly & Associates, Inc., Sebastopol, CA, **http://www.ncsa.uiuc.edu/SDG/Software/Mosaic/Docs/whats-new.html**

[49]Bryan, J. *Byte* 1995, April, pp 91 and 99. See also **http://info.cern.ch/hypertext/WWW/AccessAuthorization/Overview.html**

[50]Enterprise Integration Technologies, **http://www.eit.com/projects/s-http/**

[51]Netscape Communications Corporation, **http://home.netscape.com/info/sslref.html**

chapter seven

Designing an Internet-Accessible Database

Frank A. Tinker

In This Chapter

This chapter presents basic methods of making data available on the Internet. Some utilities and tools that can be employed in database initiation and upkeep are discussed. A working example of a bibliography of scientific papers is used to illustrate the process of implementing a database.

3223–7/96/0163$12.00/1

PERIODICALLY, MOST SCIENTISTS FIND themselves wishing that certain resources were more accessible. The resources needed can be spectra, stoichiometric relations, specific heats, bibliographies, and myriad others too numerous to mention. By the same token, you, in all probability, have information that may be vital to a fellow researcher. This chapter will give some tips on how to make what you know more available to your colleagues on the Internet.

Appropriate Internet Material

Virtually any collection of information that you once sought and compiled into a single unit can be useful to others. From a single-page list of the solid-state properties of a unique polymer to a complete library of infrared spectral lines for various types of vacuum grease, the range of subjects is limited only by your imagination.

You may know of a particularly useful resource that is currently being distributed but, for one reason or another, has become a burden to its current publisher. This was the case for our archetypal database example, "The Almost (but Never Quite) Complete Buckminsterfullerene Bibliography". The "Bucky Bib", as it is affectionately known to some, was started as a personal collection by Richard Smalley of Rice University. He, being one of the discoverers of the all-carbon molecule for which it is named, kept an informal list of fullerene papers, with occasional comments, mainly for his own reference. Soon virtually every researcher in the new field had a copy. In 1991 the Arizona Fullerene Consortium, headed by Donald Huffman, accepted the responsibility of maintaining the bibliography and making it available to other researchers.

Making information available on the Internet is essentially an exercise in publication. Consequently, the laws of copyright apply. It is neither legal nor, more importantly, ethical to publish another person's work without their consent, preferably, their written consent. Therefore, electronically scanning tables, graphs, pictures, etc., from a book or other publication and making the resulting file available on the Internet without consent is inappropriate.

In the same vein, references to other work can be extremely helpful to others. Be sure to give complete references for clarity whenever possible.

The concept of appropriateness must be distinguished from the concept of usefulness. Usefulness is more difficult to define. The best technique to determine usefulness is experimental. Make your data available, publicize it, and then monitor how frequently it is accessed. You will soon know if it is in demand.

Your Audience

Presumably the people accessing your information will be knowledgeable enough in your field or subfield to make use of it. They will not, however, all have identical computer systems or, for that matter, computer skills. Therefore, you must provide your information in a format accessible to many platforms.

The first rule of thumb is that you cannot rely on the user having a particular piece of software that may be needed to use your data. Your files should be in ASCII format if at all possible. Even if your data is formatted in a particular manner, if it is in ASCII format a simple viewer is all the user needs to read it. In many cases you may want to provide files specific to a program, like a database program. Specific programs may in fact be the best way of processing the files, but you should always try to provide one version in an ASCII format for those who do not have the specific software you use. (On Macintosh and IBMs or compatibles, saving as "text only" creates an ASCII file.)

If you want to be particularly friendly, offer the user a set of tools with which to access your data easily. You can assume that users will have their own text viewers, but a search utility or sorter that will make the data easier to cull or rearrange might be particularly useful.

Another important tip is to include instructions on how to access the data in your offering. If you have the data formatted in a particular manner, include an ASCII file explaining the formatting arrangement. Even if you do not provide software tools to exploit the formatting, those who use your data may well want to write their own.

Database Maintenance and Access Tools

A static or semistatic list of chemical properties requires little overseeing. The initial version will probably be distributed unchanged for quite a while. Other databases, such as bibliographies, require more attention. A variety of software tools can be employed in database initiation and upkeep. Many utilities are available from the Free Software Foundation, Inc. These implementations are usually tagged with the name Gnu and are freely distributable, subject to some reasonable rules. Access the Free Software Foundation, Inc., via anonymous FTP to **prep.ai.mit.edu**. (FTP is file-transfer protocol. See the appendix called Anonymous FTP.) Here are a few utilities to consider:

Editor. This is probably the tool of greatest applicability in your database maintenance arsenal. Most small or static lists can begin and end with a text editor. If you use a commercial word-processing package, remember to print your final version to a file, or save it as an ASCII file for distribution.

Database Packages. For larger lists or databases that require frequent updating, it is advisable to use a program specifically designed for database management. These packages provide convenient input, sorting, searching, and formatting capabilities that will significantly improve maintenance efficiency. Another advantage is that the database design can be changed fairly easily if it becomes necessary. Specific packages are available for certain types of databases, such as bibliographies. These packages can save you countless hours of reinventing a database design. The large number, variety, and platform availability of these types of programs makes their inclusion herein outside our general scope.

***Awk*.** This application program was "born" in the UNIX environment, but it is now available, either in source code or compiled form, for most platforms. Its purposes are to allow an automated, line-by-line input of a text file; to test for certain conditions in the current line; and to produce a screen or file output. The program can be very useful for implementing a search utility for your database and for database upkeep. The Gnu version, *gawk*, was used for the examples herein.

***Sed*.** Another program that began in UNIX and is now widely available, *sed* is short for "stream editor". It allows the editing of a text file based on a script program. This function can be useful when converting a text file from one format to another or when correcting a repeated error throughout a file.

***Grep*.** This UNIX-type utility allows a file to be searched for a particular pattern. This capability can be useful to someone wanting to search data for a particular subset of information.

`TeX`. This typesetting system, authored by Donald E. Knuth, is widely available on the Internet. `TeX` and its extension `LaTeX`, by Leslie Lamport, allow data to be readily transformed into a visually pleasing document. They also allow easy equation formatting, among many other versatile features.

`Sort`. As the name implies, `sort` is used for sorting. A number of variants of this utility are available on a range of platforms. These vary in their capabilities and command structure, so be sure to read the documentation closely.

Beyond this short list there are many other solutions to problems you may encounter in your database efforts. Search the various newsgroups and Archie, or ask your system administrator for advice. It is not usually necessary to invent a solution.

If you have studied a programming language like *C, C++, Fortran, Basic,* or *Pascal*, you can program a database solution. `Awk` and `sed` fall into the category of programming languages, but are tailored for the type of project at hand. For minor projects requiring only a small amount of coding, writing a program is a reasonable solution. For anything else, it's probably better to find a program already written by someone better qualified.

Database Design

A working example will best illustrate the process of implementing a database. A bibliography of scientific papers about fullerenes will serve as the example. A cursory examination of recent issues of *Chemical Physics Letters* provides an initial set of papers and a host of included references. Examination of these gives the following list of desirable information about each paper:

1. author(s)
2. title
3. publication date
4. submission date
5. journal name
6. journal volume
7. journal issue

We can also anticipate that other researchers will provide references for recent work not yet published; that is, they have not been accepted yet or they are in prepress form. These should also be included in the bibliography, but they need to be marked appropriately. Additional bibliographic references for the database are those listed as support in each paper obtained. However, these should be marked to inform the user that these entries have not been obtained directly and that the information might be unreliable. These two additional pieces of information will be added to the list as

8. publication status
9. confirmed valid

The next step is to determine how best to organize one record's data. The information could be typed in an acceptable bibliographic reference style with annotations:

> M. Ehbrecht, M. Faerber, F. Rohmund, V.V. Smirnov, O. Stelmakh, and F. Huisken, "CO_2-laser-driven production of carbon clusters and fullerenes from the gas phase", Chem. Phys. Lett., 214(1), 34–38, (1993) [published, valid, submitted 7/9/93]

However, this method would make searching for particular information, such as a specific submission date, or ordering by a particular field difficult. Therefore, the data for each reference is best divided into fields; that is, separate items of information. The fields will be arranged in a particular order so that programs can locate the fields in each record. The fields need to be separated from one another with a special character not used for any other purpose in the text. A vertical bar (|) will be used herein as the field separator. Finally, for multiple authors a special subfield separator is needed. A semicolon–space pair (;) will be used as the author subfield separator. A binary system will be used for the publication status and validity fields: 0 for no and 1 for yes. A database record, using these rules, will look like this:

> |M. Ehbrecht; M. Faerber; F. Rohmund; V.V. Smirnov; O. Stelmakh; F. Huisken|CO_2-laser-driven production of carbon clusters and fullerenes from the gas phase|1993|7/9/93|Chem Phys Lett |214|1|34-38|1|1|

This brings up a problem: how to tell where one record ends and the next begins. This problem can be solved by using a special separator character between records. A good choice for small records is the end-of-line character, making each line a separate record. This choice would not be good for large records, however, because editing or viewing such long lines would be difficult. A better solution is a special character that begins and ends each record. This approach allows blank lines, special "control" lines, or comment lines to be interspersed with the bibliographic data. Matched curly brackets, {}, will be used to delineate records in this example:

> {M. Ehbrecht; M. Faerber; F. Rohmund; V.V. Smirnov; O. Stelmakh; F. Huisken|CO_2-laser-driven production of carbon clusters and fullerenes from the gas phase|1993|7/9/93|Chem. Phys. Lett.|214|1|34-38|1|1}

If sorting the records by author names is desired, these names will first need to be decoded into first, last, etc. For something like A. Smith, this step is trivial. However, for a name like Herbert A. Greltsch III, the process becomes more difficult. Therefore, a consistent rule for typing author names will be needed. The rule, in this example, is Last, F.S.T...., E. The last name is spelled out completely and followed by a comma. This comma is followed by a space and the first initial followed by a period. Any additional names will be concatenated as initial–period pairs. If the author has designations after his or her name, a comma will follow the initials, then a blank space, and then the additional designations. So, "Herbert A. Greltsch III" becomes "Greltsch, H.A., III." The final, usable record format is

> {Ehbrecht, M.; Faerber, M.; Rohmund, F.; Smirnov, V.V.; Stelmakh, O.; Huisken, F.|CO_2-laser-driven production of carbon clusters and fullerenes from the gas phase|1993|7/9/93|Chem. Phys. Lett.|214|1|34-38|1|1}

The specific information in this example is immaterial. The idea is to separate the data into logical components (records), then further separate those components into subcomponents (fields), etc. In this manner, a program can be designed to easily use the component hierarchy for searching, arranging, or outputting in various formats. The number of sublevels in the hierarchy will be determined by the type of data in the records and on the level of discrimination between records that is desired.

Making the Data Readable

In this example, an editor is used to maintain the database of papers. However, instead of typing the information in a visually pleasing format, a more utilitarian formatting that will allow sorting and searching is used. Certain data (the valid and published binary fields) have even been made indecipherable to the user. The data could be distributed as is with accompanying text detailing the formatting scheme. A friendlier solution is to provide the users with tools to aid them in the use of the bibliography.

The first tool is a script (or program) that can be used to convert the data file into a more readable, ASCII format. The utility *awk* is well suited to perform the task using this script. (This chapter is not a tutorial in the *awk* language. Only a few of *awk*'s capabilities will be discussed here. To get a better idea of *awk*'s power, you should read the *awk* manual and play with the program yourself. If you do not have *awk*, ask your system administrator to get it.)

Awk reads a file, line by line, and separates the line into fields based on a field separator. The usual field separator is a space, but another character can be designated by the user. The fields are called $1, $2, etc. The entire input line is designated $0. *Awk* is invoked by typing

```
awk [-f <script.file> ] [variable=init] <input.file>
```

where the square brackets signify optional arguments. Type this line at the system prompt in UNIX or at the C prompt in an IBM-compatible PC. For Macs, check the *awk* documentation.

The first step is to write, in an editor that produces ASCII text, an *awk* script file, which is simply a file that contains one or more command clusters with the form

```
pattern { action }
```

where "pattern" is something *awk* looks for on the current line and "action" is what *awk* does if it finds that pattern. Pattern can be left blank if each line needs attention—the technique used in most of the examples that follow. Only four types of statements will be used in the following script files:

- if (conditional) statement [else statement]
- for (expression ; conditional ; expression) statement
- print expression
- variable=expression

The built-in *awk* function `substr( s, b, l )` which returns the `l`-character-long portion of string `s` starting at `b`, will also be used. For example, `substr( "I like fish", 3, 4 )` returns `like`. A variable (var) can hold information until it is needed. The types of information it can hold are either numeric (1.2345, etc.) or textual. A textual piece of information is a string. Strings are identified by having quotes around them. For example, `var1=1.2345` puts the number 1.2345 into the variable var1, and `var2="1.2345"` puts the string "1.2345" into var2. These differ in that var2 just contains a sequence of numerals and var1 contains the mathematical quantity that those numerals represent. Another example is `fish=substr( "I like fish", 3, 4 )` which puts the string "like" into the variable fish.

A conditional is simply a truth test. Only the structures

- a is equal to b (a == b)
- a is greater than b (a > **b**)
- a contains b (a ~ b)

will be used herein. These are not the only operators available; the *awk* manual lists more.

The `for` statement is used for doing something repetitively until a certain condition is met. For example, `for( i = 1 ; i < 10 ; i++ );` will put the value 1 into the variable i and then keep incrementing it (i++ is the same as i = i + 1) until it is greater than or equal to 10. That is, the `for` loop continues as long as the conditional is true.

A simple *awk* script example is

```
$1 == "I" { print $0 }
```

The $1 refers to the first word in the line (if the field separator is a space). So, `$1 = = "I"` tests if the first word is "I". With this script *awk* simply searches through a file and prints those lines that begin with the word "I". The first three words of those lines, without spaces between them, can be printed using

```
$1 == "I" { print $1 $2 $3 }
```

Putting a space between two strings in the script tells *awk* to join them together. The process is known as concatenating strings. Printing the first three words of every line in reverse order and with spaces between them is done with the following:

```
{ print $3 " " $2 " " $1 }
```

Without a pattern, all lines are operated on. The spaces are added to this action to put spaces in the output.

A slightly more complicated example combines these few *awk* statements to print each record in the example bibliography as a single line. The *awk* script looks like this:

```
#Concatenate every line to the variable whole
# When the most recent line ends with a }
# print out whole without the curly brackets.
# We always put a space between whole
# and whatever we concatenate to it. That means that
# whole always starts with a space, which is why we
# drop the first 2 characters on output instead of
# the first one.
{ whole=whole " " $0;
if(substr($0,length($0),1)=="}")
    {
    print substr(whole,3,length(whole)-3);
    whole="";
    }
}
```

(Lines beginning with # are considered comments by *awk*). If this script were saved in a file called `getrec.awk` and the bibliographic data were in a file called `data.db`, then running

```
awk -f getrec.awk data.db
```

would print each record to the screen as a single line (character-wrapped by the operating system). The record would no longer be enclosed in curly braces.

The output of `getrec.awk` is perfect for the input to another *awk* program. This is because *awk* sees a single line as a record and that's what `getrec.awk` produces.

The next *awk* program will take the output of `getrec.awk` and find all the various fields. *Awk*'s special variable *FS* will be used to designate the field separator.

```
#First set the file separator FS to the vertical
# line
BEGIN {FS="|"}
#Then the field determination is trivial
{ authors=$1; title=$2; pubdate=$3; subdate=$4; journal=$5;
volume=$6; issue=$7; pages=$8; pubbed=$9; valid=$10;
}
```

This program's only action is to fill the various variables with the fields. However, the program can be enhanced to make it output each record in a more pleasing format:

```
#First set the file separator FS to the vertical
# line
BEGIN {FS="|"}
#Then the field determination is trivial
{ authors=$1; title=$2; pubdate=$3; subdate=$4; journal=$5;
volume=$6; issue=$7; pages=$8; pubbed=$9; valid=$10;
#Start off by setting whole to a concatenation #of authors,
# title, and journal with commas in #between and title
surrounded by double quotes
    whole=authors ", \"" title "\", " journal
#Then if volume, issue, pages, and pubdate have #anything in
them, we concatenate them to whole #with formatting.
    if( length(volume) > 0 ) whole=whole ", " volume
    if( length(issue) > 0 ) whole=whole "(" issue ")"
    =if( length(pages) > 0 ) whole=whole ", " pages
    if( length(pubdate) > 0 ) whole=whole " (" pubdate ")"
#Next we check to see if we have any ancillary #information to
add. If so, we format it how we #want and concatenate it to
whole.
    if( length(subdate) || pubbed || valid )
        {
        whole=whole " [";
        if( length(subdate) > 0 )
        whole=whole "submitted " subdate ", ";
        if( pubbed ) whole=whole "published, ";
        if( valid ) whole=whole "valid, ";
        whole=substr(whole,1,length(whole)-2) "]";
        }
```

If the variable *whole* were simply printed at this point, we would almost have the desired effect. The only thing lacking is the ability to word-wrap the lines on output and to indent each record. To do this, we add

```
#Indent the record a couple of spaces
    whole=" " whole;
#Use l to keep track of our line length, beg to
# know where in whole a line begins and end to know #where
it ends.
    l = 0;
    end = 0;
    beg = 1;
#Iterate through the entire variable whole.
    for( i = 1 ; i <= length(whole) ; i++)
    {
#At the end we want to print out whatever's left
    if( i == length(whole) )
        {
        l = 75;
        end = i+1;
        }
#If the line length is greater than 72 characters, #print
out up to the last space encountered. If no #space is
```

```
encountered use character wrapping.
    if( l > 72 )
        {
        if( end <= beg ) end = i;
        print substr(whole,beg,end-beg);
        beg = end+1;
        l = 0;
        }
#Otherwise check to see if we've encountered a #space. If so,
set end to where the space is.
    else
        if( substr(whole,i,1) == " " ) end = i;
    l++;
    }
}
```

This new script file will be named `prettydb.awk`. We can now use most operating systems' capability to "chain" commands with a pipe (|) symbol:

```
awk -f getrec.awk data.db | awk -f prettydb.awk
```

and the output will be a nicely formatted version of the bibliography. As an example, here is a sample data file and how it looks after running through *prettydb.awk*:

Before

{Day, K.L.; Huffman, D.R.|Measured Extinction Efficiency Of Graphite Smoke In The Region 1200-6000A|1973|| Nature Physical Science|243||50-51|1|0}
{Rohlfing, E.A.; Cox, D.; Kaldor, A.|Production And Characterization Of Supersonic Carbon Cluster Beams|1984| 4/17/84|Journal of Chemical Physics|81||3322-3330|1|0}
{Bloomfield, L.A.; Geusic, M.E.; Freeman, R.R.; Brown, W.L.|Negative And Positive Cluster Ions Of Carbon And Silicon|1985|6/27/85|Chemical Physics Letters|121|1|33-37|1|1}
{Kroto, H.W.; Heath, J.R.; O'Brien, S.C.; Curl, R.F.; Smalley, R.E.|C60: Buckminsterfullerene|1985|9/13/85|Nature|318 ||162-163|1|0}
{Heath, J.R.; O'Brien, S.C.; Zhang, Q.; Liu, Y.; Curl, R.F.; Kroto, H.W.; Tittel, F.K.; Smalley, R.E.|Lanthanum Complexes Of Spheroidal Carbon Shells|1985|9/26/85|Journal of The American Chemical Society|107||7779-7780|1|1}

{Haymet, A.D.J.|Footballene: A Theoretical Prediction For The Stable, Truncated Icosahedral Molecule C60|1986|10/9/85|Journal of The American Chemical Society|108||319-321|1|1}

After

Day, K.L.; Huffman, D.R., "Measured Extinction Efficiency Of Graphite Smoke In The Region 1200-6000A", Nature Physical Science, 243, 50-51 (1973) [published]

Rohlfing, E.A.; Cox, D.; Kaldor, A., "Production And Characterization Of Supersonic Carbon Cluster Beams", Journal of Chemical Physics, 81, 3322-3330 (1984) [submitted 4/17/84, published]

Bloomfield, L.A.; Geusic, M.E.; Freeman, R.R.; Brown, W.L., "Negative And Positive Cluster Ions Of Carbon And Silicon", Chemical Physics Letters, 121(1), 33-37 (1985) [submitted 6/27/85, published, valid]

Kroto, H.W.; Heath, J.R.; O'Brien, S.C.; Curl, R.F.; Smalley, R.E., "C60: Buckminsterfullerene", Nature, 318, 162-163 (1985) [submitted 9/13/85, published]

Heath, J.R.; O'Brien, S.C.; Zhang, Q.; Liu, Y.; Curl, R.F.; Kroto, H.W.; Tittel, F.K.; Smalley, R.E., "Lanthanum Complexes Of Spheroidal Carbon Shells", Journal of The American Chemical Society, 107, 7779-7780 (1985) [submitted 9/26/85, published, valid]

Haymet, A.D.J., "Footballene: A Theoretical Prediction For The Stable, Truncated Icosahedral Molecule C60", Journal of The American Chemical Society, 108, 319-321 (1986) [submitted 10/9/85, published, valid]

Awk has a structure similar to the *C* programming language, which means that conversion to a dedicated, compiled *C* program is fairly straightforward. The benefit of doing so would be a significant speed improvement. However, in most cases *awk* will be sufficient.

The specific data used here is not really important to the discussion. The idea to absorb is, for larger databases, it is useful to store and manipulate the data in a different form than the one that the users will actually view. A structured text file is one way to store the data, and *awk* is one tool available to make the necessary conversions. A full-blown database program will store the data differently and provide tools to help produce a visually pleasing, distributable document. The idea is the same, the implementation is different.

You may decide, after using a text file–*awk* conversion technique, that you want a more capable database program to maintain the data. At that time you can write an *awk* script to produce a text file formatted in such a way that the database program can import it. This step will make the conversion fairly painless.

The *awk* script, `prettydb.awk`, could be altered to produce *Tex*- or *LaTex*-compatible output that will ultimately make an eye-pleasing, typeset document. This document could be kept on-file at a local print shop so that people could call and order a copy at their leisure. This is done with "The Almost (but Never Quite) Complete Buckminsterfullerene Bibliography", and the service has been widely used by many without Internet access or familiarity. *Tex* can also produce an ASCII output with centering, hanging indents, etc. The ASCII version of the "Bucky Bib" is produced this way.

Searching the Database

Formatting data into fields, as in the bibliography example, provides the basis for a program to select records based on certain search criteria. The `prettydb.awk` script can take the records so selected and output them in a more readable format. The purpose of this section is to show how one can search records using *awk*.

The *awk* match-op operator (~) evaluates to true if a certain variable or string contains another string. This means that a search for a particular author can be accomplished with the pattern

```
$1 ~ amatch { action }
```

Field number 1 is the author field. By setting the variable `amatch` to a name, *awk* will perform the desired action on all records that contain that name in the author field. The entire output routine used in `prettydb.awk` can be used as the action.

A title keyword search can be performed concurrently by using

```
$1 ~ amatch && $2 ~ tmatch { action }
```

If the author field ($1) contains the string stored in the variable `amatch`, and `(&&)` the title field ($2) contains the string stored in the variable `tmatch`, action will be performed on that record.

Execution time can be reduced somewhat by not copying the various fields into authors, title, etc. This makes the final version of the search script, called `findrec.awk`,

```
# First set our field separator to |
BEGIN {FS="|"}
# Check to see if the author and title fields #contain the
match strings (set amatch and tmatch #on the awk command line)
    $1 ~ amatch && $2 ~ tmatch {
# These are the different fields
# authors=$1; title=$2; pubdate=$3; subdate=$4; #journal=$5;
volume=$6; issue=$7; pages=$8; #pubbed=$9; valid=$10; Build the
output line using #normal bibliographic form
    whole=" " $1 ", \"" $2 "\", " $5
    if( length($6) > 0 ) whole =whole ", " $6
    if( length($7) > 0 ) whole=whole "(" $7 ")"
    if( length($8) > 0 ) whole=whole ", " $8
    if( length($3) > 0 ) whole=whole " (" $3 ")"
    if( length($4) || $9 || $10 )
        {
        whole=whole " [";
        if( length($4) > 0 ) whole=whole "submitted "
            $4 ", ";
        if( $9 ) whole=whole "published, ";
        if( $10 ) whole=whole "valid, ";
        whole=substr(whole,1,length(whole)-2) "]";
        }
# Word-wrap the output line and print
    l = 0;
    end = 0;
    beg = 1;
    for( i = 1 ; i <= length(whole) ; i++)
        {
        if( i == length(whole) )
            {
            l = 75;
            end = i+1;
            }
        if( l > 72 )
            {
            if( end <= beg ) end = i;
            print substr(whole,beg,end-beg);
            beg = end+1;
            l = 0;
            }
        else
            if( substr(whole,i,1) == " " ) end = i;
        l++;
        }
    }
```

All strings contain the null string (""), so a complete listing of all bibliographic entries can be obtained by setting `amatch` and `tmatch` to the null string before running the script. Happily, it's even easier than that because all variables in *awk* default to the null string if they are not otherwise initialized. To run the script, we'll chain together the `getrec.awk` output and the `findrec.awk` scripts using the pipe structure:

```
awk -f getrec.awk data.db | awk -f findrec.awk amatch=< > tmatch=< > .
```

where the < > are to be replaced by the desired search strings. All the references with Huffman as one of the authors can be found by running

```
awk -f getrec.awk data.db | awk -f findrec.awk amatch=Huffman
```

Of course, any combination of fields could be used in the search script to narrow the record-selection criteria. However, the example gives a working idea of how to use *awk* to find the desired records. The `alias` command in UNIX can be used to make the search process more transparent:

```
alias searchfor 'awk -f getrec.awk data.db | awk -f
    findrec.awk
    amatch=\!:1 tmatch=\!:2'
```

would allow the command string

```
searchfor Huffman
```

to be used to get a list of all the Huffman authored papers or

```
searchfor "" Lanthanum
```

to get a list of those papers with the word "Lanthanum" in the title.

If a large number of papers are listed in the bibliography, we could anticipate a large number of matches, surely more than a screenful. To handle this result simply pipe the output to the UNIX pager program called *more:*

```
searchfor Youngman | more
```

More will allow the user to page through the selected records. If a file of all matching records is desired, the input–output-redirection operators > or >> can be used. The pattern

```
searchfor "" BuckyBall > ballfile.txt
```

places all the records with titles containing the word "BuckyBall" into the text file `ballfile.txt`. To append information to an existing file, use the >> operator instead of the > operator.

Again, the actual data is immaterial. The process of searching for certain records in the database will be extremely useful both in its maintenance and in its use. Whatever tools you make to accomplish the search should be made available to your database users, along with documentation, of course.

If a database program is used to maintain the database, the search capability will be built in. However, it would still be helpful to those who use the database if they could search it easily without having to acquire and install a lot of specialized software.

Sorting the Database

Arranging the data in some useful hierarchy will make it much easier to use. Bibliographies can be sorted alphabetically by author name, title, etc., or in chronological order by publication or submission date. A table of specific heats can be sorted by temperature or alphabetically by substance, or any other reasonable manner. Trying to sort the information as it is acquired, however, is usually very time-consuming and leads to mistakes. It is better to let the computer do it.

The UNIX utility called *sort* reads input line-by-line and then arranges the lines, in order, based on a sort field. The field separator can be designated, as with *awk*, as can the particular field used for sorting. Because *sort* sees each line as a record, the `getrec.awk` script will be used to format the data into a one-record-per-line format before it is sent to *sort*. Once the database is sorted, it will be put back into multiple-line format.

To sort our records by the author field, the command

```
sort "-t|" +0 -1
```

will be used. This command specifies that a vertical line will be used as the field separator ("-t|") and that the first field, field zero (+0), is to be used as the sort field. The quotes around the "-t|" are there because the operating system would see the vertical line as a pipe symbol and not know what to do. The -1 says to ignore everything from field number 1 on in the sorting. The complete command line used to sort the data file is

```
awk -f getrec.awk | sort "-t|" +0 -1
```

The output, however, would be one record per line, which, as mentioned, is not easy to read or edit. The word-wrap portion of our `prettydb.awk` script can be used to turn the single-line records back into word-wrapped records. The *awk* script to do this will be called `ww.awk`:

```
#First reenclose the record in curly braces
{ whole="{" $0 "}"
#Then word-wrap the output line and print
    l = 0;
    end = 0;
    beg = 1;
    for( i = 1 ; i <= length(whole) ; i++)
        {
        if( i == length(whole) )
            {
            l = 75;
            end = i+1;
            }
        if( l > 72 )
            {
            if( end <= beg ) end = i;
            print substr(whole,beg,end-beg);
            beg = end+1;
            l = 0;
            }
        else
            if( substr(whole,i,1) == " " ) end = i;
        l++;
        }
    }
}
```

Now, the command line

```
awk -f getrec.awk data.db | sort "-t|" +0 -1 | awk -f
    ww.awk > datasort.db
```

will input the data file `data.db`, sort it, and output it to the new file `datasort.db`. The data can be sorted alphabetically by title with the command string

```
awk -f getrec.awk data.db | sort "-t|" +1 -2 | awk -f
    ww.awk > tsort.db
```

Here are the results of these sorts on the data used earlier:

Author Sort

{Bloomfield, L.A.; Geusic, M.E.; Freeman, R.R.; Brown, W.L.|Negative And Positive Cluster Ions Of Carbon And Silicon|1985|6/27/85| Chemical Physics Letters|121|1|33-37|1|1}
{Day, K.L.; Huffman, D.R.|Measured Extinction Efficiency Of Graphite Smoke In The Region 1200-6000A|1973|| Nature Physical Science| 243||50-51|1|0}
{Haymet, A.D.J.|Footballene: A Theoretical Prediction For The Stable, Truncated Icosahedral Molecule C60|1986| 10/9/85|Journal of The American Chemical Society|108|| 319-321|1|1}
{Heath, J.R.; O'Brien, S.C.; Zhang, Q.; Liu, Y.; Curl, R.F.; Kroto, H.W.; Tittel, F.K.; Smalley, R.E.|Lanthanum Complexes Of Spheroidal Carbon Shells|1985|9/26/85|Journal of The American Chemical Society|107||7779-7780|1|1}
{Kroto, H.W.; Heath, J.R.; O'Brien, S.C.; Curl, R.F.; Smalley, R.E.|C60: Buckminsterfullerene|1985|9/13/85|Nature| 318||162-163|1|0}
{Rohlfing, E.A.; Cox, D.; Kaldor, A.|Production And Characterization Of Supersonic Carbon Cluster Beams|1984| 4/17/84|Journal of Chemical Physics|81||3322-3330|1|0}

Title Sort

{Kroto, H.W.; Heath, J.R.; O'Brien, S.C.; Curl, R.F.; Smalley, R.E.|C60: Buckminsterfullerene|1985|9/13/85| Nature|318||162-163|1|0}
{Haymet, A.D.J.|Footballene: A Theoretical Prediction For The Stable, Truncated Icosahedral Molecule C60|1986| 10/9/85|Journal of The American Chemical Society|108|| 319-321|1|1}
{Heath, J.R.; O'Brien, S.C.; Zhang, Q.; Liu, Y.; Curl, R.F.; Kroto, H.W.; Tittel, F.K.; Smalley, R.E.|Lanthanum Complexes Of Spheroidal Carbon Shells|1985|9/26/85|Journal of The American Chemical Society|107||7779-7780|1|1}
{Day, K.L.; Huffman, D.R.|Measured Extinction Efficiency Of Graphite Smoke In The Region 1200-6000A|1973|| Nature Physical Science| 243||50-51|1|0}
{Bloomfield, L.A.; Geusic, M.E.; Freeman, R.R.; Brown,

W.L.|Negative And Positive Cluster Ions Of Carbon And Silicon|1985|6/27/85| Chemical Physics Letters|121|1|33-37|1|1}
{Rohlfing, E.A.; Cox, D.; Kaldor, A.|Production And Characterization Of Supersonic Carbon Cluster Beams|1984|4/17/84|Journal of Chemical Physics|81||3322-3330|1|0}

Awk and `sort` provide a huge capability for structuring and formatting the data. One can search, sort, and print (using *prettydb.awk)* the results using one pipe after another. A dedicated database program would have built-in sorting capabilities that would make `sort` and *awk* unnecessary for maintenance, but the user might still benefit from your `sort`- and *awk*- based tools.

Making Data Available on the Internet

Once your data has been compiled, manipulated, and sorted and various versions have been produced (raw, pretty ASCII, TeX, etc.), it should be made available to those who would find it useful in their work. The other chapters of this book provide details of how to implement various techniques of Internet access to your data. You and your system administrator will need to determine which techniques are best for you, your data, and your site.

Probably the most useful access to your data is via anonymous FTP (file-transfer protocol). This type of access is an implementation of an FTP system at your site that will allow users to log in as "anonymous" or "FTP" and then give their electronic-mail (e-mail) addresses as passwords. This type of access is easily ported to the World Wide Web and Gopher so that a wide range of users can get your database offering. Your system administrator will need to implement the FTP site for you but you can get a wealth of ideas from the World Wide Web (Chapter 6) and Gopher (Chapter 5) information in this book.

Monitoring the number of accesses to your FTP site can give you feedback on how well your database is received. To get a list of all the users accessing your site, you can use the UNIX command string

```
last ftp | grep ^ftp > ftp.use
```

This will put the list of sites accessing yours into a file named `ftp.use`. You can then use *awk* and `sort` on this file to get daily, monthly, and regional statistics on those who access the site.

Another useful technique for distributing database information, provided it is not too large, is via a list server. Such a server will respond to persons sending e-mail messages to request copies of the data. The interested party can even request to be placed on a subscription list that automatically sends updates when they become available. The data is sent via e-mail, which limits the prudent size of the data file distributed in this manner but increases the number of potential users. Chapter 4 presents detailed tips on implementing such a server.

Another, less used, technique of offering your data on the Internet is to allow people to log onto your system and then interactively search the database. The "Bucky Bib" is available in this manner and is often accessed this way. The overhead for providing this service, however, is usually prohibitive. If you feel that such a technique is worth the effort, check with your system administrator for help. You can also try to work out an arrangement with a library to make your database available this way.

Final Words

The tips and techniques herein are not the only ones at your disposal in implementing, maintaining, and distributing a database to your peers on the Internet. They are workable, however, and are adaptable to many situations and contexts. You are encouraged to seek improvements to or replacements for these techniques by examining the huge volume of lists, tables, notes, and graphs found in countless archives across the Internet. Spend a week "surfing" the Net looking for examples and you will, no doubt, find more examples than could possibly be chronicled here.

A conspicuous paradox of the Internet is its growth as an effective resource pool in spite of, or perhaps because of, the scarcity of regulation. Such a trend has been, and always will be, due the efforts of those individuals willing to share their knowledge and expertise for no other reason than a sense of responsibility. If your aim is to join this club, you will be warmly welcomed. Good luck.

chapter eight

Electronic Conferencing

Thomas C. O'Haver

In This Chapter

This chapter presents a description of electronic conferences and tells how they differ from traditional conferences. The technical requirements are outlined, and guidelines are given for holding an electronic conference. The early experiments in on-line conferences are described.

3223–7/96/0185$12.00/1

THE TRADITIONAL TECHNOLOGIES for public scientific discourse and communication include print (books and journals) and transportation (meetings and conferences). Both are characterized by a long history and by a very advanced state of development and integration into modern culture. In recent years these traditional technologies have been joined by newly developed electronic technologies based on video and computer networks. These newer technologies are already being adopted and adapted by the broader scientific community, as described in other chapters in this volume, and in some cases are even being used to replace older technologies. In some corporations and research organizations, for example, electronic conferencing using sophisticated real-time, two-way video technology is beginning to replace conventional meetings.

Electronic conferencing uses communication technology to allow people to "convene" electronically, without physically traveling to a common location. This is part of a general trend toward increasing reliance on "telecommuting", that is, using telephones, fax machines, and computer networks rather than physical travel to conduct business. The greater efficiency of electronic alternatives and substantial savings in travel time and fuel and paper consumption are motivating forces for this trend.

Two different kinds of technology can be used for electronic conferencing: video-based and document-based. Video conferencing seeks to duplicate as closely as possible the experience of a conventional conference by showing real-time video images of speakers, discussions, and groups of people. Document-based conferencing uses electronic document and file transfer over networks to deliver written papers and supporting documents and to enable interactive discussions in writing. In a sense, document-based conferencing is really closer to electronic publishing, but with an added element of interactive discourse.

The objectives of engaging in scientific discourse and communication at scholarly conferences, by whatever technologies may be used, include exchanging ideas, describing ongoing projects, sharing experiences, presenting research results, establishing priority to new ideas, enhancing personal and institutional reputation and prestige, reaching decisions, asking questions, and forming and maintaining contacts and relationships with other scholars. In evaluating the impact of newer technologies on the way scholarly conferences might be conducted, we must ask whether these objectives can be met more fully or at lower cost than with conventional means.

Technical Requirements for Effective Electronic Conferencing

An important step in planning an electronic conference is to decide on the level of technology to be used, taking into consideration what is available to the intended participants. If the conference is a small one within the confines of one organization, then it may be possible to choose one of the commercial video-conferencing systems and to provide all the necessary hardware, software, and training to all participants. On the other hand, if the organizers want to attract a large number of participants from various organizations and locations, and if they expect the participants to use their own local hardware, then they have to search for a technology that is already widely available. A tradeoff between accessibility to a large audience and a high technological level will always be necessary.

In general, large-scale international electronic conferences have up to now relied on standard Internet services based on TCP/IP (transmission control protocol/Internet protocol), which are discussed in other chapters in this book, rather than dedicated conferencing systems or real-time video conferencing. Text-based store-and-forward technologies, mostly using electronic mail (e-mail) and Usenet News, have been used to handle topical discussions and announcements. (A store-and-forward technology is a communication service in which messages sent by one person to another individual or to a group are stored until the recipients access them at a later time.) Information resource technologies such as file-transfer protocol (FTP), Gopher, or the World Wide Web (WWW) have been most commonly used to handle the distribution of the papers, graphics, instructions, software, etc. The participants will need to have accounts on their own organizational host systems and will have access to either host-based clients or local clients for the required services. (A client is software that runs on the computer receiving the information.) On the other hand, if a dedicated conferencing system is to be used, then the organizers will have to provide client software as part of the packet sent to all paid registrants.

In either case, the participants must make a certain investment in learning how to use the selected software and communication services. A substantial advantage of using standard Internet services rather than a proprietary dedicated conferencing system is that the standard Internet services have many other uses and the investment in learning them will be repaid in other applications. The reminder of this chapter will concentrate on the use of general-purpose Internet services.

Conference registration can be carried out by either postal or e-mail, and registration fee payment can be handled by either postal mail or some form of electronic fund transfer. Restriction of conferences to paid registrants can be ensured by giving passwords to paid registrants.

Document Delivery

In a document-based electronic conference, various supporting documents, such as schedules, abstracts, papers, "slides" (graphics), and instructions, are delivered to participants electronically, typically by means of e-mail, FTP, Gopher, or the WWW on "Web pages". For most users, e-mail is limited to text documents or text-encoded documents. On the other hand, FTP and Gopher can handle binary documents that can include graphical elements, provided that the participants can obtain or are provided with the necessary client or file-viewing software.

The most universal format is, of course, ASCII text, but it is also the least satisfactory for chemistry and other scientific documents that make extensive use of non-ASCII characters. Binary file formats of popular word-processing programs are a possibility, but so many different versions of word processors are in common use that compatibility is a major problem. Another alternative is rich text format (RTF), a platform-independent standard for formatted text that supports fully formatted text and embedded graphics. The latest versions of virtually all commercial word-processing programs support RTF. Commercial document-exchange products such as Adobe's `Acrobat` can neatly solve the cross-platform compatibility problems; however, they require special-purpose document-preparation software that would have to be supplied to all authors of presented papers, as well as reader software that would have to be supplied to all participants.

A most attractive method of complex document delivery at present is the HTML (HyperText markup language) format employed by the World Wide Web, especially when using graphical Web clients such as `Mosaic` or `Netscape`. A substantial advantage is that Web pages can integrate formatted text with embedded graphics, can have HyperText-style links that can automatically call up another document, and can handle digitized sound and video with the appropriate "helper" applications. It is even possible to extend the suite of file formats and specialized viewer applications recognized and automatically employed by graphical Web clients.

Minor disadvantages of this approach are that it takes some effort to get documents into the required format and that only participants with directly networked personal computers will be able to take full advantage of the multimedia possibilities. The HTML format employed by the WWW is not so rich in format control and complex typography as a modern scientific word processor, but it is evolving rapidly and has the advantage that the Web browsers that interpret it are widely available at no cost for all the popular Windows, Macintosh, and UNIX workstation platforms. For chemistry applications it is fortunate that the latest version of the HTML specification (HTML 3.0) supports subscripts and superscripts.

Discussion and Conversation

An important part of any conference is the discussion between participants that takes place either privately in the hallways or publicly in the meeting rooms after each talk. In the text-based conference, the discussion can be either synchronous, in which case conferees participate at the same time but in different places, or asynchronous, in which case conferees participate at different times of the day over the conference period, relying upon electronic media to store messages and forward them to each participant.

Synchronous text-based discussion, in which messages are transmitted immediately to recipients, can be carried out by means of Internet Relay Chat (IRC) or by the " rooms" or "forums" of commercial bulletin board systems (BBSs). Asynchronous text-based discussion, in which messages are saved until recipients accesses them, can be carried out by means of e-mail, `listserv` lists, Usenet newsgroups, or the internal e-mail facilities' BBS.

The synchronous mode is closer to the dynamics of a traditional physical meeting, but asynchronous conferencing allows people with different schedules or even in different time zones to participate. Matthew Rapaport, a researcher who has studied many kinds of computer-mediated communications, reports that "All researchers and developers in the field agree that most real work is accomplished asynchronously, where the store-and-forward capabilities of the software increase the communication potential of the medium across time zones and personal schedules".[1] Even more significant, experience has shown that the relaxed time frame of asynchronous discussion results in writing that is more thoughtful and scholarly in tone than the more immediate synchronous formats.

An advantage of dedicated conferencing systems is that they usually include decision-support facilities for voting and polling of participants, with automatic tallying of results. Limited voting and polling can also be carried out by e-mail and newsgroup posting, but results will have to be manually processed.

An attractive alternative to the use of e-mail list servers or newsgroups is to use WWW browsers that support fill-in-the-blank "forms" (most of the recent ones do). An asynchronous discussion can be carried out by using a form designed to allow individual participants to type their responses into a scrolling text field and then press a "submit" button to submit their contributions. A custom-designed forms-processing shell script or program running on the host computer can collect these contributions and add them to a discussion page that grows in length as the discussion proceeds, and each contribution is marked with the name of the contributor and the time and date. Even multiple simultaneous discussions can be handled with this approach using a separate discussion page for each session. The primary advantage of this approach is that the entire conference can be carried out using only one network tool. Scripts may even be developed to support voting and polling of participants; votes could be registered by clicking on check boxes, radio buttons, and pop-up lists, and the results could be automatically tallied.

Guidelines for Creating an On-line Conference

The following is a suggested sequence of steps for planning and conducting an on-line conference. With suitable modifications, this procedure may be adapted to different types and sizes of on-line conferences.

1. Assemble a conference committee to decide on the scope, topic, structure, and time frame of the conference.
2. Arrange for a sponsoring organization or institution. An on-line conference is much less costly to conduct than a conventional conference, but a small budget to pay for honoraria, student helpers, and other incidental expenses may be useful. Include a representative of the sponsoring organization on the conference committee, if desired.

3. Make arrangements for a host computer system to handle the distribution of papers and the discussion. This task can most likely be done through a cooperating college or university. The conference committee should include at least one technical person, preferably someone located at the site of the host computer system, who is able and authorized to handle the technical aspects (writing instructions, helping authors format their papers, uploading files, setting up and maintaining the on-line discussion facilities, etc.)
4. If multiple sessions on different subtopics are planned, locate suitable session leaders who are willing to organize and preside over the various authors and send out invitations, describe the topic and time frame, and provide instructions for the submission of papers.
5. If the conference is to include contributed papers, distribute a Call for Papers to the print media most likely to be read by the prospective participants and to any related electronic media (list servers, newsgroups, and BBSs) describing the conference topics and time frame and providing instructions for the submission of titles and abstracts to the appropriate session leaders. Participants may submit titles, abstracts, and registration forms by e-mail or by HTML form, if the conference is to be limited to those with access to the WWW.
6. Make a general announcement of the conference as for the Call for Papers, including instructions for registration to the conference. Invite representatives from the appropriate print media to attend and monitor the conference.
7. If a conference registration fee is required, appoint a treasurer to handle checks and receipts.
8. As submitted titles and abstracts arrive, have the session leaders acknowledge all submissions, accept suitable papers, reject unsuitable ones, and report the status of their sessions to the conference committee.
9. Distribute detailed instructions for authors to guide them in converting their papers and graphics (figures) into a suitable format for electronic transmission (e.g., HTML with graphics in graphics interchange format, or GIF).
10. The conference committee and the session leaders decide on a conference schedule, that is, over what time periods the various papers are to be discussed. Distribute the schedule to the authors, and inform them that they are expected to be available during that period and that they should check their e-mail or newsgroup several times per day during their discussion period.

11. Distribute a conference program electronically to all registrants via the conference discussion facility. This program should contain the schedule of sessions, a complete set of instructions for conference participants, an evaluation form (if desired), and pointers to sources of software for viewing the papers. Alternatively, floppy disks containing the required viewing software could be distributed to registered participants via postal mail, at additional expense.
12. Inform participants that they are expected to access and read the papers before the scheduled discussion periods.
13. Declare the conference open. The session leaders introduce each author and paper, possibly allowing the author to make the first comment, and then "opening the floor" for questions and discussion. The session leaders should monitor the discussion to keep it on track and on time.
14. Inform participants that technical questions concerning conference access, e-mail, communication software, viewing software, etc., should be sent directly to the conference technical person, not to the conference itself. As needed, the conference committee may distribute important announcements to the entire conference.
15. If desired, schedule a general discussion session or follow-up session after the papers' discussion sessions. Plans for publication of the conference proceedings and for future on-line availability may be announced. Future conferences may be discussed.
16. Distribute evaluation forms to participant discussion facilities and ask that they be returned via private e-mail to a designated address.
17. If desired, the conference discussion facility may be disabled to prevent its random use between conferences, or it may be left to announce the next conference.
18. The conference committee should prepare a report to the sponsoring organization describing the conference and its evaluation by participants and authors.

The First On-line Conferences

The CHEMCONF '93 Experiment

During the summer of 1993, the Committee on Computers in Chemical Education of the American Chemical Society's Division of Chemical Education sponsored the world's first on-line electronic conference for chemistry educators, entitled "Applications of Technology in Teaching Chemistry".[2–6] Donald Rosenthal, of Clarkson University, and I organized it. Fifteen papers were "presented", on topics ranging from the applications of computer animation and simulation of chemical reactions to using classroom networks to foster cooperative learning. The idea of CHEMCONF was to use global network facilities to conduct a conference on the topic of chemistry education that would be accessible to chemistry educators around the world at no cost.

CHEMCONF was organized much as a conventional research conference, with a series of invited or contributed papers and a discussion period following each of the papers in which the participants are invited to ask questions of the authors and discuss among themselves any issues raised by the papers. A conference schedule was distributed to all participants, specifying the date that each paper would be available for (electronic) distribution and the dates that paper would be open for discussion. The conference schedule, abstracts of the papers, and instructions were distributed to all participants via the `listserv` list well in advance of the start of the conference. Each paper was made available 1 week before discussion was scheduled to begin.

The discussion was carried out on the conference `listserv`. All participants were encouraged to participate in these discussions. The discussions were scheduled sequentially. Participants were able to switch the flow of discussion messages off and on at any desired time, so they could listen only to those discussions that interested them.

Because the CHEMCONF conference was planned in 1992, before the WWW was fully operational, we adopted a procedure for the distribution of graphics and photographs in electronic form based on formats and network services that were widely available at that time: ASCII text for the text and GIF (graphics interchange format) for the graphics, which is a multiplatform standard for color bit-map graphics.

Documentation for the process for creating graphics (for authors) and for viewing graphics (for participants) and all necessary software and instructions were made available to conference participants on an Internet server that is accessible via Telnet, FTP, Gopher, or WWW clients. For participants limited to e-mail communication, GIF graphics and other binary files were made available in text-encoded form and encoders–decoders were also made available. Some of the papers described IBM-PC and Macintosh software developed by the authors; several of those authors provided working copies of their software, which were installed on the host system for participants to download and try.

A total of 450 chemistry educators from 33 countries participated, generating the equivalent of more than 1000 pages of discussion concerning the papers and related topics. After the conference, participants filled out and returned detailed evaluation forms. Many respondents reported that the electronic format permitted wider participation, greater interaction, and more thoughtful, in-depth discussions, compared to a conventional meeting. The obvious advantages of cost and time savings were commonly mentioned, but beyond that, many participants expressed the feeling that the quality of the discussion was improved compared to conventional conferences and that a deeper and more detailed examination of topics was facilitated because of the freer time frame and emphasis on author–listener interaction.

The most common complaint was the great volume of e-mail that was generated in a short time. Many participants were simply not accustomed to using e-mail in a "serious" and intensive manner demanded by a conference. Another commonly cited dissatisfaction was the frequency of off-the-subject comments and small-talk. Some effort from a moderator is necessary to keep the discussion on track. Some participants missed not being able to see the other people. In retrospect, it would have been technically feasible to have included digitized photographs of the authors with the papers.

Even without face-to-face contact, however, some participants reported that they had made many valuable contacts and felt that they had "gotten to know" some of the other participants through their writing. A few people even found that the relative anonymity of the network was an advantage, helping them overcome the psychological barrier of speaking in public.

Among the conference participants were three chemistry professors from Gallaudet University, the renowned university for the hearing-impaired in Washington, DC. Hearing-impaired participants are not at all handicapped in a document-based, on-line conference of this sort. Another participant was a quadriplegic chemist for whom conventional conferences are very difficult.

Although the first CHEMCONF conference is now concluded, the papers, graphics, discussion streams, evaluations, software, instructions, and related material are still available on-line, via WWW at **http://www.inform.umd.edu/EdRes/Faculty_Resources_and_Support/Chem Conference/** or via Gopher or FTP at **inform.umd. edu**, in the path

/EdRes/Faculty_Resources_and_Support/ChemConference. The conference list server is **CHEMCONF@umdd. umd.edu**.

INCINC '94, the International Chemometrics Internet Conference

The success of the first conference experiment inspired other successful chemistry-related on-line conferences organized along similar lines.[7] INCINC '94, the first International Chemometrics Internet Conference, was held from September 26 to November 18, 1994, and was organized by Barry Wise of by Pacific Northwest Laboratory and sponsored by the North American Chapter of the International Chemometrics Society (NAmICS) and by Elsevier, publisher of *Chemometrics* and *Intelligent Laboratory Systems*.

The topic sessions were Chemometrics and Education; Chemometrics: Philosophy, History, and Directions; Data Rectification/Validation; Chemometrics in Dynamic Systems; Nonlinear Alternative to Neural Networks; Chemometrics in Chromatography; QSAR/Chemometrics in Molecular Modeling; Second-Order Calibration; and Three-Way Analysis. The conference archive site was **ftp.pnl.gov** (path **/emsl/incinc94/outgoing**), and the discussion list server was **INCINC94@UMDD.UMD.EDU**.

Two significant technical innovations of INCINC '94 were (1) the use of a WWW server (**http://www.emsl.pnl.gov:2080**) to deliver papers, and (2) the use of the *listserv* TOPICS facility to allow participants to selectively subscribe to one or more parallel discussion streams.

The First Electronic Computational Chemistry Conference

The First Electronic Computational Chemistry Conference (ECCC) was held from November 7 through November 18, 1994, and extended for an additional 2 weeks due to popular demand.[8,9] The ECCC was organized by Steven Bachrach and sponsored by Northern Illinois University and the Henry and Camille Dreyfus Foundation. The conference was open to all aspects of computational chemistry The papers spanned a very broad definition—from ab initio, semiempirical, and molecular mechanics calculations, to drug design and QSAR, to molecular dynamics and Monte Carlo simulations, to the use of computers in chemical education.

The ECCC was designed to emulate as closely as possible the format of the familiar conference poster session. All papers were viewed exclusively via the WWW. Abstract submission was handled by a forms-based HTML page or by e-mail. All 75 papers were submitted in HTML. The authors were responsible for preparing their manuscripts in HTML; the conference instructions included many helpful pointers to Web resources for preparing HTML documents. Most papers were served by the NIU Chemistry WWW server (**http://hackberry.chem.niu.edu:70/0/ECCC/homepage.html**), but about 20 were served locally by the authors. The ECCC home page contained links to all of the papers. Discussion among the more than 300 participants was handled by a list server. Participants were urged to include the paper number of the paper they were discussing in the subject line so that other participants could filter the mail. Furthermore, the e-mail discussions were converted to HTML pages using *hypermail*, and the subject line was used to sort the mail messages. A forms-based HTML page was used to survey the participants on their response to the ECCC and electronic conferencing in general.

The proceedings of the ECCC, including the papers and e-mail discussions, are available on CD-ROM for the Macintosh and Windows platforms. The CD-ROM contains all the papers and e-mail discussions that occurred during the conference as well as WWW browsers and viewers.

The Electronic Conference on Trends in Organic Chemistry

The Electronic Conference on Trends in Organic Chemistry (ECTOC) was held from June 12 through July 7, 1995.[10,11] Papers and posters were accessed exclusively through the World Wide Web and were discussed using e-mail. The conference was organized by Henry S. Rzepa, Jonathan Goodman, and Chris Leach, and was sponsored by the Royal Society of Chemistry Glaxo Group Research (Greenford), Daylight (Santa Fe), International Society of Heterocyclic Chemist, and SmithKline Beecham Pharmaceuticals. The ECTOC conference papers were fully indexed using an algorithm that facilitates finding relevant information.

Several papers included supplemental information, such as molecular structure drawings and three-dimensional molecular coordinates, in standard forms that can be visualized by participants using special-purpose viewing or "helper" programs. The popular WWW browser

Netscape has the ability to detect HyperText links to specialized data types and to launch a specified program (called a "helper") to view that data. This facility is commonly used to view digital video clips and to hear digitized sound files. The ECTOC organizers explained how to extend the list of recognized special data types to include Adobe *Acrobat* files, molecular coordinates in pdb (Protein Data Bank) format, and structure drawings in ISIS/Draw format. After installing the required helper programs and configuring *Netscape* accordingly, the participant need only click on the HyperText link and the data are automatically downloaded and visualized by the appropriate helper program.[12,13] The ECTOC Web page is located at **http://www.ch.ic.ac.uk/ectoc/**.

The Second Electronic Computational Chemistry Conference

The Second Electronic Computational Chemistry Conference (ECCC-2) took place from November 1 through November 30, 1995. This conference was sponsored by the Camille and Henry Dreyfus Foundation through a grant in the Chemical Informatics Program. As in previous Web-based conferences, extensive use was made of specialized chemical data types to represent chemical structures, molecular coordinates, X-ray structure data, and spectroscopic data. An innovation of ECCC-2 was that it was held entirely on the Web.

Unlike previous electronic conferences, where discussions of the papers was handled using e-mail, discussions of papers in ECCC-2 were held with Web technology, using a user interface designed specifically for the conference. Each paper was listed by title. Selecting a paper would take the user to a page containing the abstract and a list of messages pertaining to that paper. Each user was able to list all messages or only the new messages, send a new message, or respond to a previous message, all by using a few mouse clicks. Further information on ECCC-2 can be obtained from the Web page at **http://hackberry.chem.niu.edu/ECCC2**.

Future On-line Conferences

The future clearly promises continued growth in electronic conferencing. Conferences are appearing in ever-more-diverse disciplines and using the most current technological advances.

The Electronic Conference on Heterocyclic Chemistry

The Electronic Conference on Heterocyclic Chemistry (ECHET96) will be the second electronic conference in the ECTOC series on trends in organic chemistry, this time with the focus on heterocyclic chemistry. Discussion of submitted articles will occur from June 24 through July 22, 1996, on the Internet using the World Wide Web and electronic mail as access mechanisms. Scientific organization will be under the auspices of the International Society for Heterocyclic Chemistry and the Royal Society of Chemistry Perkin Division, Heterocyclic Group. ECHET96 will feature full-color structure diagrams, "clickable" or editable reaction schemes, rotatable three-dimensional molecular images, index searches of conference proceedings, contributed discussions by participants, a conference "molecular hyperglossary", up-to-date access statistics for articles, photographs submitted by participants, and other novel features. Most likely, ECHET96 will be abstracted by the Chemical Abstracts Service.

Contributions will take the form of a number of invited keynote articles, together with submitted articles and conference posters. Articles will be regarded as equivalent to journal articles or communications; they will be subjected to the peer-review process; and they will also be duly published in CD-ROM form. We will also encourage accepted posters to be included on the conference CD-ROM.

Preliminary abstracts of articles or posters (or their URLs) should be posted to the conference editors at **echet96-contrib@ic.ac.uk** by April 26, 1996, at the latest. Full conference information and details of how to submit contributions will be available at the following World Wide Web locations using a browser such as *`Mosaic`* or *`Netscape`*: **http://www.ch.ic.ac.uk/ectoc/echet96/** or by sending an e-mail message to **listserver@ic.ac.uk** with the single-line message: `information echet96`.

ChemConf '96: New Initiatives in Chemical Education

ChemConf '96, an on-line symposium entitled "New Initiatives in Chemical Education", is sponsored by the Division of Chemical Education of the American Chemical Society and will be held from June 3 to July 19, 1996. Don Rosenthal of Clarkson University and I are the conference organizers. The papers will be made available on a WWW site, and the discussion will be carried out via a *`listserv`* discussion list (**chemconf@umdd.umd.edu**). Further information on the program is available on the ChemConf '96 home page at URL **http://www.inform.umd.edu:8080/EdRes/Faculty_Resources_and_Support/ ChemConference/ChemConf96/Home.html**

Acknowledgments

My deepest thanks to the Computer Science Center of the University of Maryland, College Park, for their generous support of the host system requirements for the CHEMCONF '93 and INCINC '94 conferences.

References

[1]Rapaport, Matthew *Computer-Mediated Communications;* Wiley: New York, 1991.

[2]Borman, Stu "On-line Chemical Education Meeting May Be Harbinger of the Future", *Chem. Eng. News*, February 15, 1993, p 25.

[3]Borman, Stu "Benefits of Electronic Conferences Cited by On-line Meeting 'Attendees'", *Chem. Eng. News*, October 11, 1993, p 26.

[4]"Good Chemistry at Electronic Conference", *Science (Washington, D.C.)* **1993,** *261*, 1271.

[5]Bodow, Steve "New Formula for Conferences", *Wired*, November 1993, p 30.

[6]O'Haver, Thomas C. "Meeting in Cyberspace", *The World & I*, February, 1994, p 218.

[7]Krieger, J. H.; Illman, D. L. *Chem. Eng. News,* December 12, 1994, pp 29–40.

[8]*Proceedings of the First Electronic Computational Chemistry Conference, Nov. 7–Dec. 2, 1994;* Bachrach, S. M.; Boyd, D. B.; Gray, S. K.; Hase, W.; Rzepa, H. S.; Eds.; ARInternet: Landover, MD, 1995.

[9]Bachrach, S. M. *J. Chem. Inf. Comput. Sci.* **1995,** *35,* 431.

[10]Rzepa, H. S. *Trends Anal. Chem.* **1995**, in press.

[11]Krieger, J. H. *Chem. Eng. News,* August 21, 1995, p 35.

[12]Rzepa, H. S.; Whitaker, B. J.; Winter, M. J. *J. Chem. Soc. Chem. Commun.* **1994,** 1907.

[13]Casher, O.; Chandramohan, G.; Hargreaves, M.; Leach, C.; Murray-Rust, P.; Sayle, R.; Rzepa, H. S.; Whitaker, B. J. *J. Chem. Soc., Perkin Trans 2* **1995,** 7.

part two

Specifically for Chemists

chapter nine

Electronic Lists for Chemists

Jan K. Labanowski, Charles F. Bender, and Alejandro Pisanty

In This Chapter

This chapter contains information about electronic lists that focus on chemistry. Information on the following lists is given:

3223–7/96/0203$37.00/1

MANY ELECTRONIC LISTS discuss topics related to chemistry. Several excellent documents that describe chemistry-related services and resources are available via the Internet. These documents enumerate electronic lists, electronic archives, World Wide Web sites, file-transfer protocol (FTP) sites, Gopher sites, important electronic-mail addresses, and other sources of information. The well-known document, "Chemistry-Related Resources on the Internet", is maintained by Gary Wiggins (**wiggins@indiana.edu** or **wiggins@indiana.bitnet**) and is available from many electronic archives. The most recent version of this document can be retrieved by either an anonymous FTP (**cica.cica. indiana.edu**) or a gopher (**gopher.cica.indiana.edu**) from the CCIIM (Clearing House for Chemical Information Instructional Materials) site. It can be found there by using the following directory path: `CICA Research (pub)/chem/resource.010`. Other valuable compilations of various chemical resources available on the Internet can be found on the anonymous FTP site **leon.nrcps.ariadne-t.gr** and also on the World Wide Web site **http://bionmr1. rug.ac.be/chemistry/overview.html**.

The compilation that follows resulted from searching various archives and from a query posted to several chemical newsgroups. The compilation represents the status of the lists in October 1994. This collection is far from complete. For one thing, it is difficult to decide what is and what is not related to chemistry. Omissions may also be due to the fact that many discussion lists are private and do not seek publicity. In such cases, we simply did not find any references to them. Finally, we did not include lists that are run as a for-fee service.

A modified version of this compilation will be kept in the archives of the Computational Chemistry List and can be retrieved by anonymous FTP from **ccl.osc.edu** from the directory `/pub/chemistry/info/electronic-lists/chemistry-related-lists` or by sending the message:

```
select chemistry
get info/electronic-lists/chemistry-related-lists
quit
```

to the address MAILSERV@ccl.osc.edu.

For some lists we had more information than for others. Where available we give the editor, average traffic on the list, number of subscribers as of October 1994, location of archives, and subscription procedures or the name of the list-management software.

The subscription and posting procedures are usually similar between different list managers. Each entry gives specific instructions to subscribe, unsubscribe, and post ,or an administrative address that accepts subscription and unsubscription requests and a distribution address for posting messages to all subscribers.

Remember to place the subscription or unsubscription request or command within the body of your message (not on the `Subject:` line) because most automatic list subscribers and unsubscribers do not analyze the `Subject:` header line. It is generally considered unethical to post to a list if you are not an active subscriber. Besides, you may miss the most interesting responses, which will go to the list. To enforce this rule many electronic lists will not accept messages from nonsubscribers. It is also important to check the rules of the list before you post. They should give you a scope of the list. Post only material that is directly related to the list topic. Thousands of electronic lists and newsgroups are operating, and most likely, there is a forum for each valid comment or question.

The letter case in e-mail addresses is not important; lowercase and uppercase letters can be mixed to enhance readability. However, traditionally, addresses referring to UNIX machines are usually in lowercase, and the Bitnet addresses are in uppercase. Also, the letter case of commands usually does not matter, but the file names or Web URLs (uniform resource locators) should be typed exactly as shown. To be on the safe side, always enter commands and their components in the letter case shown.

Name: AMALGAM—Amalgam / mercury / health
Description: Questions about amalgam, mercury, and health are discussed.
Subjects: amalgam, mercury, health
Editor: Siegfried Schmitt, e-mail **je04@ibm3090.rz.uni-karlsruhe.de**
To Subscribe: Send the following message in the first line of the body: `SUB AMALGAM firstname lastname` to the address **LISTSERV@listserv.net**
To Unsubscribe: Send the message `SIGNOFF AMALGAM` to the address **LISTSERV@listserv.net**
To Post: Send mail to **AMALGAM@listserv.net**
Archives: All messages from 1992 to the present are archived by year and are retrievable from the list server by any subscriber.

Name: AMBER—Amber users discussions
Description: A mail reflector for users and people interested in Amber. A major part of the traffic is the bug fix announcements.

Subjects: amber programs, force field
Editor: Bill Ross, Department of Pharmaceutical Chemistry 1913, University of California, San Francisco, CA 94143–0446, fax (415) 476–0688, e-mail **amber-request@cgl.ucsf.edu**
Comments: Unmoderated. 20 messages/month. 280 subscribers. Not archived. List type: UNIX.
Administrative address: **amber-request@cgl.ucsf.edu** or **amber-request@ucsfcgl.bitnet**
Distribution address: **amber@cgl.ucsf.edu** or **amber@ucsfcgl.bitnet**

Name: ANALYSIS-L—Analytical Science
Description: Discussion of all aspects of analytical science, including techniques and methodologies. This list is not just for the discussion of analytical chemistry but for the wider subject area of analytical science.
Editor: Andy Packham, e-mail **mcnajp@fs4.in.umist.ac.uk**
To Subscribe: Send the message `SUBSCRIBE Analysis-L` to **maiser@fs4.in.umist.ac.uk**
To Unsubscribe: Send the message `UNSUBSCRIBE Analysis-L` to **maiser@fs4.in.umist.ac.uk**
To Post: Send mail to **Analysis-L@fs4.in.umist.ac.uk**

Name: ANNEAL—Simulated Annealing
Description: Covers theory and application of simulated annealing, which is a computer algorithm for performing combinatorial optimization. It is used for a variety of problems, including VLSI placement, network optimization, protein folding, and logic minimization. Members are typically researchers in the field.
Subjects: thermodynamics, organic chemistry, optimization, protein configuration
Editor: Dan Greening, Novell, 1601 Saratoga-Sunnyvale Rd, Suite 100, Cupertino, CA 95014, phone (408) 973–8081 ex 313, fax (408) 973–0989, e-mail **greening@sti.com**
Comments: Unmoderated, 5 messages/month, no archives, 250 subscribers. List type: UNIX. Please investigate the algorithm on your own before posting questions to this list.
Administrative address: **anneal-request@cs.ucla.edu**
Distribution address: **anneal@cs.ucla.edu**.

Name: APCHEM—Astrophysical Chemistry
Description: The Astrophysical Chemistry Group (ACG) of the Royal Society of Chemistry, which is also affiliated to the Royal Astronomical Society, held its inaugural meeting in April 1994. The purpose of this group is to make a bridge between observational astronomy and atomic

and molecular physics and chemistry. Meetings highlight topics of outstanding importance in astronomy in areas in which atomic and molecular physics and chemistry can make a major contribution. Important functions of the ACG are to make astronomers aware of what chemists can deliver, to make chemists aware of what astronomers seek, and to encourage collaborative projects between astronomy and chemistry.

Subjects: astrophysical chemistry, interstellar chemistry, planetary chemistry, atmospheric chemistry, physical chemistry, theoretical chemistry, computational chemistry

Editors: Bernd R. Eggen, University of Exeter, e-mail **apchem-request@mailbase.ac.uk**. David Field, University of Bristol, e-mail **D.Field@Bristol.ac.uk**

Comments: Unmoderated. Low volume. 100 participants. Archived.
List type: MAILBASE.
Administrative address: **mailbase@mailbase.ac.uk**
Distribution address: **apchem@mailbase.ac.uk**

Archives: Available via WWW **http://mailbase.ac.uk/** and anonymous FTP **mailbase.ac.uk** in the directory `/pub/lists/apchem`. See also `/pub/mailbase`.

Name: APPLSPEC—Applied spectroscopy

Description: The Society of Applied Spectroscopy (SAS) mailing list

To Subscribe: Send the message `SUBSCRIBE APPLSPEC your_name` to the address **LISTSERV@uga.cc.uga.edu**

To Unsubscribe: Send the message `SIGNOFF APPLSPEC` to the address **LISTSERV@uga.cc.uga.edu**

To Post: Send mail to **APPLSPEC@uga.cc.uga.edu**

Name: BIOSCI/BIONET—Biological Science Groups and Lists

Description: BIOSCI is a series of freely accessible electronic communication forums (i.e., electronic bulletin boards or newsgroups) for use by biological scientists worldwide. No fees are charged for the service. The system is intended to promote communication between professionals in the biological sciences. There are more than 60 newsgroups and corresponding electronic lists on Bionet ranging from typical biological subjects like cell biology, genetics, and parasitology to interdisciplinary fields including molecular modeling, NMR structure determination, and protein crystallography.

Subjects: virtually all areas of biology, biochemistry, and biophysics

Editor: Each list or newsgroup has its own leader. Bionet is managed by David Kristofferson (**kristoff@net.bio.net**).

To Subscribe: Once you choose the particular Bionet list and you know its LISTNAME, subscribe by sending the message `subscribe LISTNAME` to the address**biosci-server@net.bio.net**

To Unsubscribe: Send the message `unsubscribe LISTNAME` to the address **biosci-server@net.bio.net**

To Post: Send your message to the address **some-list@net.bio.net** where the some-list is the appropriate list identifier

Comments: Bionet is an excellent, professionally run resource, with clear and extensive instructions on its use and topics. The archives of Bionet are extensive. Start by sending the message `info faq` to the address **biosci-server@net.bio.net**. You may also use anonymous FTP or **gopher net.bio.net** and retrieve tutorials on the use of Bionet and a description of Bionet lists and newsgroups. Some Bionet lists are moderated.

Name: BUCKY—Fullerene News

Description: A weekly list of the latest titles on C_{60} and related fullerenes, fullerites, and molecular and solid-state derivatives.

Subjects: fullerenes, fullerites

Editor: Jack Fischer, 3231 Walnut Street, Philadelphia PA 19104–6202, phone (215) 898–6924, e-mail **fischer@lrsm.upenn.edu**

To Subscribe: Send the message `SUBSCRIBE` to the address **bucky@sol1.lrsm.upenn.edu**

To Unsubscribe: Send a message `UNSUBSCRIBE` to **bucky@sol1.lrsm.upenn.edu**

To Post: Place the single word `SUBMISSION` on the first line and follow it by up to 10 lines of message. Send it to **bucky@sol1.lrsm.upenn.edu**. A successful submission will be acknowledged with an automatic response.

Comments: Unmoderated. 27 messages/month. 1200 subscribers. Archived.

Archives: Send the message `INTRO` (note capital letters) to **bucky@sol1.lrsm.upenn.edu** for more information. Send a message to **bucky@sol1.lrsm.upenn.edu** with the keyword titles or biblio to get the latest set of submissions or the full Bucky bibliography, respectively. Also maintained at Arizona State U. Physics Department as anonymous FTP (and eventually WWW/Mosaic).

Name: C2-L—Cerius2 Users Discussions

Description: This list is for users of Cerius2. This is an MSI Scientific Support Bulletin Board mailing list intended to provide an informal forum for users of MSI products to discuss various aspects of the programs. The list is not designed to be an official channel for product support; all requests for technical product support should continue to go through the Scientific Support hotline.

Subjects: molecular mechanics, molecular dynamics, proteins, nucleic acids

Editor: Robert B. Funchess, Molecular Simulations Inc., 16 New England Executive Park, Burlington, MA 01803–5297, phone (617) 229–9800 ex 202, fax (617) 229–9899, e-mail **bobf@msi.com**, WWW **http://www.msi.com/~bobf/bobf.html**

Comments: Unmoderated, 10 messages/month, 100 subscribers, Archived.
List type: LISTPROC.
Administrative address: **listproc@msi.com**
Distribution address: **c2-l@msi.com**

Archives: Archived monthly. Anonymous FTP **server.msi.com**, WWW **http://www.msi.com/**, or via e-mail to **listproc@msi.com** (send index and then request needed files by sending `get filename` command to **listproc@msi.com**).

Name: CAChe—CAChe Users Discussions.

Description: Intended to enhance communication between academic users of computer modeling for teaching. In particular, the focus is on applications of the modeling tools from CAChe Scientific in the undergraduate curriculum. Items appropriate for this forum include classroom experiences, course outlines, modeling projects, sample results, bug reports, new software announcements, requests for help, and trouble-shooting tips.

Subjects: modeling, thermodynamics, organic chemistry, physical chemistry, inorganic chemistry, CAChe Scientific

Editor: James Currie, Chemistry Department, Pacific University, Forest Grove, OR 97116, phone (503) 357–6151, fax (503) 359–2933, e-mail **jim_currie@unixmail.pacificu.edu**

Comments: Unmoderated. No subscription restrictions. 5 messages/month. 55 subscribers. No archives. List type: MAJORDOMO.
Administrative address: **majordomo@pacificu.edu**
Distribution address: **cache@pacificu.edu**

Name: CHARMM-BBS—CHARMM Users Discussions

Description: To electronically connect users—of all levels of expertise—of the CHARMM molecular mechanics package. Comprises users from academia and industry and is a free service. Charmm-bbs is *strictly* a secular environment; it is *not* maintained by MSI (commercial version of CHARMm) or by Harvard (including the Karplus group).

Subjects: molecular mechanics and dynamics, modeling, classical mechanics

Editor: Donovan N. Chin, Harvard University, 12 Oxford Street, Cambridge, MA 02138, phone (617) 495–9435, fax (617) 495–9857, e-mail **dchin@emperor.harvard.edu**

Comments: Unmoderated. No subscription restrictions. 5 messages/month. 230 subscribers. Archived.
Administrative address: **charmm-bbs-request@emperor.harvard.edu**
Distribution address: **charmm-bbs@emperor.harvard.edu**

Archives: Messages are archived but not available on line at this time (mid-1995). To obtain selected previous messages, send a request stating which messages you need to the address **charmm-bbs-request@ emperor.harvard.edu**

Name: CHEMCOM—Discussion of ACS courses
Description: For discussion of the ACS course called ChemCom and any auxiliary topics. Questions on lessons, labs, ideas, etc. A list for teachers around the world who are teaching Chemistry in the Community, the course designed by the American Chemical Society and taught in high schools. This list is used to post questions and lesson ideas (what works and what doesn't).
Editor: Thomas L. Gath, Cortland High School, Cortland NY 13045, phone (607) 753–6061, fax (607) 753–6115,
e-mail **TGATH@OCMVM.CNYRIC.ORG**
Comments: Unmoderated. 100 messages/month. 125 subscribers. Not archived, but digest available via **LISTSERV@UBVM.CC.BUF-FALO.EDU**. List type: LISTSERV.
Administrative address: **LISTSERV@UBVM.CC.BUFFALO.EDU** or **LISTSERV@UBVM.BITNET**
Distribution address: **CHEMCOM@UBVM.CC.BUFFALO.EDU** or **CHEMCOM@UBVM.BITNET**

Name: CHEM-COMP—Computational Chemistry List
Description: For the discussion of all aspects of computational chemistry
Subjects: computational chemistry
Editor: Andrew Sherwood, **cstajs@staffs.ac.uk**
Comments: Unmoderated. No subscription restrictions. Very low traffic. 220 subscribers. Archived. List type: MAILBASE.
Administrative address: **mailbase@mailbase.ac.uk**
Distribution address: **chem-com@mailbase.ac.uk**
Archives: Archived automatically. Access via e-mail (`MAILBASE GET` command; send the message `send mailbase user-guide` to **mailbase @mailbase.ac.uk** for instructions on e-mail retrieval), anonymous FTP **mailbase.ac.uk**, gopher **mailbase.ac.uk**, WWW **http://mailbase.ac.uk/**

Name: CHEMCONF—CHEMistry CONFerence
Description: Devoted to the operation of on-line computer conferences in any area of academic chemistry research and education. Specific conferences on particular topics will be conducted from time to time; dates and topics will be announced via this list and in the chemistry print journals (e.g., *Chemistry and Engineering News*). This list is intended to serve as the main mechanism for the distribution of the texts of papers invited by or submitted to the conference and for the discussion of those papers among participants and authors. Members of the chemistry research and education community who wish to propose topics for future confer-

ences should contact T. C. O'Haver, Department of Chemistry and Biochemistry, University of Maryland, College Park, MD 20742 (**to2@umail.umd.edu**).

Subjects: chemistry, education, conferences

Editor: T. C. O'Haver, Department of Chemistry and Biochemistry, University of Maryland, College Park, MD 20742, e-mail **to2@umail.umd.edu**

Comments: Unmoderated. No subscription restrictions. 1–2 messages/month between conferences. Hundreds of messages/month during a conference. 500 subscribers. Archived. List type: LISTSERV.
Administrative address: **listserv@umdd.umd.edu**
Distribution address: **chemconf@umdd.umd.edu**

Archives: Send the message `INDEX CHEMCONF` to **listserv@umdd.umd.edu** to obtain the file list. Then retrieve the file by sending the message `GET CHEMCONF filename` to the address **listserv@umdd.umd.edu**

Name: CHEME-L—Chemical Engineering List

Description: Unedited forum on the role of chemical engineering in a changing technology and world economy and its adaptation to those changes; new research trends in industry and academia; public and federal financial support availability; public image of the chemical engineering profession; role of the professor. Information items on available opportunities for postdoctoral fellows and faculty or graduate students on leave. Collaborative research. Free forum on all research and education aspects of chemical engineering.

Subjects: chemical engineering, industrial chemistry, engineering profession, engineering education, social impact of chemical engineering

Editor: Raoul Miranda, Department of Chemical Engineering, University of Louisville, Louisville, KY 40292–0001, phone (502) 852–0191, fax (502) 852–6355, e-mail **r0mira01@ulkyvm.louisville.edu**

Comments: Unmoderated. 60 messages/month. 500 subscribers. Not archived. List type: LISTSERV.
Administrative address: **LISTSERV@PSUVM.PSU.EDU**
Distribution address: **CHEME-L@PSUVM.PSU.EDU**

Name: CHEMECOL—Chemical Ecology

Description: Discussion of topics related to chemical ecology

To Subscribe: Send a message to **listserv@iqm.unicamp.br** with the following line: `subs chemecol Your Personal Name`

To Unsubscribe: Send a message to **listserv@iqm.unicamp.br** with the following line: `signoff chemecol`

To Post: Send mail to **chemecol@iqm.unicamp.br**

Name: CHEMED-L—Chemical Education
Editor: Bill Halpern, e-mail **WHALPERN@UWF.BITNET**
Comments: Unmoderated, No subscription restrictions. 50 messages/month. 900 subscribers. Archived. List type: LISTSERV.
Administrative address: **LISTSERV@UWF.BITNET**
Distribution address: **CHEMED-L@UWF.CC.UWF.EDU**
Archives: Send the message `INDEX CHEMED` to **LISTSERV@UWF.CC.UWF.EDU** to receive the list of files. Then for each file to be retrieved send the message `GET CHEMED-L filename` to **LISTSERV@UWF.CC.UWF.EDU**

Name: CHEMIC-L—Chemistry in Israel List
Description: A forum for announcing academic activities (seminars, meetings, job offers, etc.) in Israel
Subjects: chemistry
Editor: **CHEMISTL@TAUNIVM.BITNET**
Comments: Unmoderated. No subscription restrictions. Up to 20 messages/month. 180 subscribers. List type: LISTSERV.
Administrative address: **LISTSERV@VM.TAU.AC.IL**
Distribution address: **CHEMIC-L@VM.TAU.AC.IL**
Archives: Send the message `INDEX CHEMIC-L` to the address **LISTSERV@VM.TAU.AC.IL** to get file list. Retrieve a file by sending the message `GET CHEMIC-L filename` to the address **LISTSERV@VM.TAU.AC.IL**

Name: CHEMIND-L—Chemical Structure Indexing
Description: A discussion list that serves as a forum for the discussion of questions relating to indexing and searching chemical structure information. News and questions about new developments in indexing and searching techniques, new structure-searchable databases and systems (on-line, in-house, or on CD–ROM), and developments in existing databases are suitable topics. The full range of indexing techniques from textual codes to graphical structures is covered with no restriction on the range of types of structure—specific structures of all types (organic, inorganic, organometallic, polymers, macromolecules, etc.) and Markush structures typically found in patents.
Subjects: chemical structure indexing, searching, databases
Editor: Bill Town (who is also the list owner), Derwent Information Ltd., Derwent House, 14 Great Queen Street, London, England WC2B 5DF, phone +44 71 344 2800, fax +44 71 344 2911,
e-mail **btown@derwent.co.uk**

Comments: Unmoderated. Archived. List type: LISTSERV.
Administrative address: **listserv@derwent.co.uk**
Distribution address: **CHEMIND-L@DERWENT.CO.UK**

Archives: Available via e-mail. To retrieve messages for a given month send the message `GET CHEMIND-L LOGyymm` to **listserv@derwent.co.uk**. For example, for July 1994 messages: `GET CHEMIND-L LOG9407`

Name: CHEMISTRY—Computational Chemistry List

Description: Not restricted to any particular method or software, but topics like quantum chemistry, molecular mechanics and dynamics, molecular modeling, and computer-aided molecular design dominate the discussions.

Subjects: quantum chemistry, molecular modeling, chemical kinetics, molecular graphics, molecular design

Editor: Jan Labanowski, Ohio Supercomputer Center, 1224 Kinnear Road, Columbus, OH 43212, phone (614) 292–9279, fax (614) 292–7168, e-mail **jkl@osc.edu** or **JKL@OHSTPY.BITNET**

Comments: Unmoderated. 10 messages/day. 2000 subscribers. Archived.
Administrative address: **chemistry-request@ccl.osc.edu**
Distribution address: **chemistry@ccl.osc.edu**

Archives: Updated on line by incoming messages. Anonymous FTP **ccl.osc.edu** in directory: `/pub/chemistry`.
WWW **http://ccl.osc.edu/chemistry.html**, gopher **ccl.osc.edu 73,**
e-mail, send message

```
HELP CHEMISTRY
HELP ENCODERS
HELP GET
HELP
```

to **MAILSERV@ccl.osc.edu** for detailed info on how to retrieve files from the archive.

Name: CHEMISTRYTM—Chemistry Telementoring

Description: An international telementoring discussion group for students in high school and college and their teachers whose interest (avocation, coursework, or career) lies in the area of chemistry. The group fosters the exchange of ideas and information on all topics related to chemistry. Chemistrytm is an open (unmoderated) list intended to include students from every country connected to Bitnet and the Internet.

Subjects: chemistry

Editor: Oliver Seely, Department of Chemistry, California State University, Dominguez Hills, Carson, CA 90747, e-mail **oliver@dhvx20.csudh.edu**

Comments: Unmoderated, 5–10 messages/month, archives to be made accessible, 80 subscribers. List type: UNIX.
Administrative address: **chemistrytm-request@dhvx20.csudh.edu**
Distribution address: **chemistrytm@dhvx20.csudh.edu**

Name: CHEM-MOD—Modeling in Chemistry
Description: For discussion of the modeling aspects of computational chemistry.
Subjects: computational chemistry, modeling
Editor: Andrew Sherwood, **cstajs@staffs.ac.uk**
Comments: Unmoderated. Very low traffic. 160 subscribers. Archived. List type: MAILBASE.
Administrative address: **mailbase@mailbase.ac.uk**
Distribution address: **chem-mod@mailbase.ac.uk**
Archives: Archived automatically. Access via e-mail (`MAILBASE GET` command-send the message `send mailbase user-guide` to **mailbase@mailbase.ac.uk** for instructions on e-mail retrieval), anonymous FTP **mailbase.ac.uk**, gopher **mailbase.ac.uk**,
WWW **http://mailbase.ac.uk/**

Name: CHMINF-L—Chemical Information Sources
Description: Provides a forum where people can discover new reference tools and databases of interest to chemists and also find answers to reference questions of a chemical nature. It is in many respects a virtual chemistry reference library.
Subjects: chemistry, chemical information science, reference questions
Editor: Gary Wiggins, Chemistry Library, Indiana University, Bloomington, IN 47405, phone (812) 855–9452, fax (812) 855–6611,
e-mail **wiggins@indiana.edu** or **wiggins@indiana.bitnet**
Comments: Unmoderated. No subscription restrictions. 150 messages/month. 1050 subscribers. Archived. List type: LISTSERV.
Administrative address: **LISTSERV@IUBVM.BITNET** or **LISTSERV@IUBVM.INDIANA.EDU**
Distribution address: **CHMINF-L@IUBVM.BITNET** or **CHMINF-L@IUBVM.INDIANA.EDU**
Archives: Monthly log files are maintained at Indiana University. Send the command `GET CHMINF-L LOGyymm` (where yy is year and mm is month, expressed numerically) to **LISTSERV@IUBVM.BITNET** or **LISTSERV@IUBVM.INDIANA.EDU**. A gopher archive is available from the chemistry gopher at the University of California at Davis, URL **gopher://gopher-chem.ucdavis.edu:70/1/1/Email_Archive/ChmInf-L**

Name: CICOURSE—Chemical Information Courses on the Internet
Description: Provides a forum in which people can teach and learn about chemical information sources in all forms. Open to both students and instructors. Participants may choose to limit the information received to

one or more of the 10 topics: lectures, slides, readings, errors, discussions, homework, exams, pedagogy, evaluations, and miscellany, or by default receive all messages.

Subjects: chemistry, chemical information sources, on-line searching, databases, reference sources, chemical communication

Editor: Gary Wiggins, Chemistry Library, Indiana University, Bloomington, IN 47405, phone (812) 855–9452, fax (812) 855–6611, e-mail **wiggins@indiana.edu** or **wiggins@indiana.bitnet**

Comments: Unmoderated. No subscription restrictions. 30 messages/month. 300 Subscribers. Archived. Proposals for workshops, courses, etc., should be sent to the list owner, Gary Wiggins. List type: LISTSERV. Administrative address: **LISTSERV@IUBVM.BITNET** or **LISTSERV@IUBVM.INDIANA.EDU**
Distribution address: **CICOURSE@IUBVM.BITNET** or **CICOURSE@IUBVM.INDIANA.EDU**

Archives: The list of archive files can be obtained by sending the message `GET CICOURSE FILELIST` to **LISTSERV@IUBVM.BITNET** or **LISTSERV@IUBVM.INDIANA.EDU**

Name: CORROS-L—Corrosion and Degradation

Description: Concerned with corrosion and degradation of materials. It covers technical questions, general information, and the notification of conferences, job vacancies, etc.

Subjects: corrosion, electrochemistry, materials

Editor: Bob Cottis, Corrosion and Protection Centre, UMIST, P.O. Box 88, Manchester M60 1QD, UK, phone 44 161 200 4843, fax 44 161 200 4865, e-mail **R.A.Cottis@umist.ac.uk**

Comments: Unmoderated. No subscription restrictions. 10 messages/month. 100 subscribers. Archived. List type: LISTSERV. Administrative address: **LISTSERV@IB.RL.AC.UK**
Distribution address: **CORROS-L@IB.RL.AC.UK**

Archives: Sent the message `INDEX CORROS-L` to the address **LISTSERV@IB.RL.AC.UK** to get a list of files. Retrieve files by sending the message `GET CORROS-L LOGyymm` (where yy is year and mm is month) to **LISTSERV@IB.RL.AC.UK**. For more information send the message `INFO DATABASE` to **LISTSERV@IB.RL.AC.UK**

Name: CZE-ITP—Electrophoresis

Description: Discussion list on capillary electrophoretic methods, including zone electrophoresis, micellar electrokinetic chromatography, isotachophoresis, and isoelectric focusing.

Subjects: capillary electrophoresis, zone electrophoresis, isotachophoresis, isoelectric focusing, micellar electrokinetic chromatography

Editor: Vladislav Dolnik, Institute of Analytical Chemistry, Academy of Sciences of the Czech Republic, Veveri 97, CZ-611 42 Brno, Czech Republic, phone 42–5–7268145, fax 42–5–4121–2113, e-mail **vdolnik@ipm.cz**

Comments: Unmoderated. Up to 10 messages a month. 100 subscribers. Archived. List type: LISTSERV.
Administrative address: **listserv@vm.ics.muni.cz**
Distribution address: **CZE-ITP@vm.ics.muni.cz**

Archives: Anonymous FTP **ftp.muni.cz** in the directory `pub/muni.cz/cze-itp`

Name: DIBUG— Discover, Insight, Biosym Users Group

Description: An unmoderated mailing list for discussion of Biosym products, maintained by Biosym software users and researchers. Expected topics of discussion include, but are not limited to, queries about how to use features of the software and how to solve scientific problems using the software. Although this list is administered by and for Biosym users, it is expected that the vendor monitors the list, and perhaps responds to issues that may arise.

Subjects: general aspects of molecular modeling, molecular dynamics, ab initio and semiempirical methods, meeting announcements, NMR and X-ray crystallography data evaluation, aspects of force field usage and application, efficiency of use

Editor: Reinhard Doelz, Biocomputing, Biozentrum der Universitaet Basel, Klingelbergstr.70, CH 4056 Basel, Switzerland, phone 41 61 267 2247, fax 41 61 267 2078, e-mail **doelz@ubaclu.unibas.ch**, WWW **http://beta.embnet.unibas.ch**

Comments: Unmoderated. 50 messages/month. 450 subscribers. Archived.
Administrative address: **dibug-request@comp.bioz.unibas.ch**
Distribution address: **dibug@comp.bioz.unibas.ch**

Archives: WWW, gopher, anonymous FTP **bioftp.unibas.ch**

Name: FORENS-L—Forensic Medicine and Sciences

Description: An unmoderated discussion list dealing with forensic aspects of anthropology, biology, chemistry, odontology, pathology, psychology, serology, toxicology, criminalistics, and expert witnessing and presentation of evidence in court.

Subjects: forensic science

Editor: List owners: M. Yasar Iscan, e-mail **Iscan@acc.fau.edu**, and Ronald K. Wright, e-mail **RKW@MedExam.FtL.FL.US**. List manager: Ralph P. Carpenter, e-mail **Ralpho@acc.fau.edu**

To Subscribe: Send the message `SUBSCRIBE FORENS-L Firstname Lastname` to the address **MAILSERV@ACC.FAU.EDU** or **MAILSERV@FAUVAX.BITNET**

To Unsubscribe: Send the message `UNSUBSCRIBE FORENS-L` to the address **MAILSERV@ACC.FAU.EDU** or **MAILSERV@FAUVAX.BITNET**
To Post: Send your message to **FORENS-L@ACC.FAU.EDU** or **FORENS-L@FAUVAX.BITNET**
Comments: Unmoderated

Name: GA Digest—Genetic Algorithm List
Description: Posting of news, opinions, questions, and discussions of issues interesting to the genetic algorithms community. Also announcements for related conferences, seminars, and books.
Subjects: genetic algorithms, evolutionary programming, evolution strategies, genetic programming (these are all known as "evolutionary algorithms")
Editor: William M. Spears (Current Moderator), e-mail**GA-List-Request@aic.nrl.navy.mil**
Comments: Moderated. 4 messages/month. 2700 subscribers. Archived. Administrative address: **GA-List-Request@aic.nrl.navy.mil** Distribution address:**GA-List@aic.nrl.navy.mil**
Archives: **ftp.aic.nrl.navy.mil** is an anonymous FTP site. It contains papers by the AIC staff, code, conference announcements, and back issues of the digest. Should be up on WWW soon.

Name: GA-MOLECULE—Genetic Algorithms in Chemistry
Description: Brings together researchers who want to solve problems in chemistry using genetic algorithms. Unlike GA-list, this is an unmoderated mailing list. Since its inception just after ICGA 1991, it has had very low traffic, so don't be surprised if you don't hear anything for several weeks!
Subjects: genetic algorithms, molecular-structure prediction
Editor: Tom Ngo, Interval Research Corporation, 1801-C Page Mill Road, Palo Alto, CA 94304–1216, phone (415) 354–3616, fax (415) 354–0872, e-mail **ngo@interval.com**
To Subscribe: Send the message `help` to **ga-molecule-request@interval.com**. In principle send `subscribe ga-molecule` or (if you wish to be subscribed with another address) `subscribe ga-molecule other-address` to the address **ListMgr@interval.com**
To Unsubscribe: Send the message `unsubscribe ga-molecule` or (if you were subscribed with address different than your current account) `unsubscribe ga-molecule other-address`) to **ListMgr@interval.com**
To Post: Send your posting to **ga-molecule@interval.com**
Comments: Unmoderated. No subscription restrictions. Low traffic. 130 subscribers. Archived.

Archives: Send the message `index ga-molecule` to **ListMgr@interval.com** to obtain list of available files. Then send the message `get ga-molecule filename` to retrieve the specific file.

Name: HIRIS-L—HIgh Resolution Infrared Spectroscopy
Description: Discussions on high-resolution IR spectroscopy
Subjects: IR spectroscopy
Editor: **G1WUDD32@ICINECA.BITNET** or **G1WUDD32@ICINECA.CINECA.IT**
Comments: Unmoderated. 150 subscribers. List type: LISTSERV.
Administrative address: **LISTSERV@VM.CINECA.IT**
Distribution address: **HIRIS-L@VM.CINECA.IT**

Name: HYDROGEN—Hydrogen as Fuel
Description: A forum for the exchange of information and ideas pertaining to the production, storage, and use of hydrogen as an alternative fuel. The purpose of this list is to promote a better understanding of the concepts, terminology, materials, processes, and issues relating to hydrogen. Users are welcome from universities, government, and industry, and are encouraged to post all pertinent news, information, research, references, seminar and conference announcements, product and service announcements, related procurements, and the general discussion of topics.
Subjects: hydrogen, chemical energy storage, energy, alternative energy, fuel, renewable energy, fusion
To Subscribe: Send the message `SUB HYDROGEN Firstname Lastname` to **LISTSERV@URIACC.URI.EDU**. You will need to respond to the confirmation message for the registration to be accepted.
To Unsubscribe: Send the message `UNSUB HYDROGEN` to **LISTSERV@URIACC.URI.EDU**
To Post: Send your message to **HYDROGEN@URIACC.URI.EDU**
Comments: Unmoderated. 50 messages/month. 125 participants. Archived.

Name: HYPERCHEM—Hyperchem Users Group
Description: Designed primarily for, but not limited to, HyperChem users. Any and all scientific and technical issues related to the use of HyperChem are appropriate for discussion, and we hope that it will be a valuable resource for HyperChem users. News and developments concerning HyperChem will also be posted to this list.
Subjects: HyperChem, ChemPlus, HyperNMR, computational chemistry, product announcements
Editor: Graham Hurst, Hypercube Inc, 7-419 Phillip Street, Waterloo, Ontario, Canada N2L 3X2, phone (519) 725–4040, fax (519) 725–5193, e-mail **hurst@hyper.com**

To Subscribe: Send the message `subscribe hyperchem` or the longer form `subscribe hyperchem ADDRESS` (replace ADDRESS with your e-mail address) to the address **hyperchem-request@hyper.com**
To Unsubscribe: Send the message `unsubscribe hyperchem` or the longer form `unsubscribe hyperchem ADDRESS` (replace ADDRESS with the e-mail address you are subscribed with) to **hyperchem-request@hyper.com**
To Post: Send e-mail to **hyperchem@hyper.com**
Comments: Unmoderated. Up to 30 messages/month. 400 subscribers. No archives.

Name: ICS-L—International Chemometric Society
Description: The discussion group of the International Chemometrics Society, a not-for-profit scholarly society that is concerned with the development and application of mathematical and statistical methods for the analysis of chemical data. The purpose of the list is to serve as a medium for discussion and information exchange among our members and any other interested scholars. Discussion should center on methods, algorithms, software, hardware platforms, data sets, meetings and conferences, and current research trends and applications.
Subjects: Chemistry, chemometrics, statistics, mathematical methods
Editor: T. C. O'Haver, Department of Chemistry and Biochemistry, University of Maryland, College Park, MD 20742, e-mail **to2@umail.umd.edu**
Comments: Unmoderated. No subscription restrictions. 5 messages/month. 220 subscribers. Archived. List type: LISTSERV. Administrative address: **listserv@umdd.umd.edu** Distribution address: **ICS-L@umdd.umd.edu**
Archives: Send the message `INDEX ICS-L` to **listserv@umdd.umd.edu** to get the list of files. Then retrieve files by sending the message `GET ICS-L filename` to the address **listserv@umdd.umd.edu**

Name: INTERF-L—Interfacial Phenomena
Description: Intended for physicists, chemists, biologists and engineers whose research interests lie within the broad area of physics and chemistry of interfacial phenomena, in particular, (1) liquids at interfaces, wetting and adhesion phenomena, capillarity; (2) complex and macromolecular fluids: structure, phase transitions, and interfaces; (3) thermodynamics and hydrodynamics of polymer melts and solutions, microemulsions, colloids, emulsions and foams, suspensions, surfactants in solutions, membranes, amphiphilic monolayers and bilayers, gels, liquid crystals, and porous media. The INTERF-L Interest Group will provide the following functions: (1) increase the communication between members of this interdisciplinary community in Israel; (2) announce

seminars, conferences, scientific programs, grants, and other items of common interest.

Subjects: interfacial phenomena, surface physics, complex fluids, soft condensed matter, polymers, colloids, emulsions, foams, liquid crystals, amphiphiles, suspensions, membranes, gels, wetting, adhesion, surfactants, macromolecular fluids

Editor: David Andelman, School of Physics, Tel Aviv University, Ramat Aviv 69978 Israel, phone +972–3–6407239, fax +972–3–6422979, e-mail **andelman@taunivm.tau.ac.il**

Comments: Moderated. 1 message/month. 105 subscribers. No archives. List type: LISTSERV.
Administrative address: **listserv@taunivm.tau.ac.il**
Distribution address: **interf-l@taunivm.tau.ac.il**

Name: INTERFACE-L—Instrument Interfacing

Description: To encourage and facilitate the discussion of all aspects of computer-instrumentation interfacing. The discussion is not limited to any one computer type or operating system.

Subjects: computation, instrumentation, interfacing

Editor: Andy Packham, Department of Instrumentation and Analytical Science, University of Manchester, Institute of Science and Technology, Manchester, PO Box 88, M60 1QD, UK, phone 061 200 4899, fax 061 200 4911, e-mail **mcnajp@fs4.in.umist.ac.uk**

To Subscribe: Send the message `SUBSCRIBE Interface-L Firstname Lastname` to the address **MAISER@FS4.IN.UMIST.AC.UK**

To Unsubscribe: Send the message `UNSUBSCRIBE Interface-L` to the address **MAISER@FS4.IN.UMIST.AC.UK**

To Post: Send your message to the address **Interface-L@FS4.IN.UMIST.AC.UK**

Comments: Unmoderated. 10 messages/month. 100 subscribers. Not archived, though some files are available on anonymous FTP server **fs4.in.umist.ac.uk**. List type: MAISER.
Administrative address: **MAISER@FS4.IN.UMIST.AC.UK**
Distribution address: **Interface-L@FS4.IN.UMIST.AC.UK**

Name: METALLURGY-L—Metallurgical Engineering

Description: A discussion list for all aspects of metallurgical engineering, including (but is not necessarily limited to) mineral processing, extractive metallurgy, hydrometallurgy, pyrometallurgy, metals refining, alloying, welding, casting, and metallography. It is mainly intended both for students and professional metallurgists. Topics include technical discussion, meeting announcements, and other materials that the members feel are appropriate.

Subjects: metallurgy

Editor: Timothy Eisele, Department of Metallurgical and Materials Engineering, Michigan Technological University, Houghton, MI 49931, e-mail **tceisele@mtu.edu**

Comments: List type: MAJORDOMO.
Administrative address: **majordomo@mtu.edu**
Distribution address: **metallurgy@mtu.edu**

Name: MICROSCALE-L—Microscale Chemistry

Description: To answer questions about microscale chemistry. Run by the National Microscale Chemistry Center at Merrimack College.

Subjects: microscale, organic, inorganic, general, high school, elementary school

Editor: Zvi Szafran, Dean of Science and Engineering, Merrimack College, North Andover, MA 01845, phone (508) 837–5328, fax (508) 837–5029, e-mail **Zszafran@merrimack.edu** or **DrZviS@aol.com**

To Subscribe: Send the one-word message `subscribe` to the address **microscale-l-request@merrrimack.edu**

To Unsubscribe: Send the one-word message `unsubscribe` to the address **microscale-l-request@merrrimack.edu**

To Post: Send your message to **microscale-l@merrimack.edu**

Comments: Unmoderated. 200 messages/month. Archives will be soon made available on WWW **http://nmcc.merrimack.edu**

Name: MMODINFO—MacroModel Users Discussion Group

Description: An e-mail exploder that allows MacroModel users and computational chemistry researchers to exchange information and experience.

Subjects: MacroModel software

Editor: Douglas A. Smith, Concurrent Technologies Corporation, 1450 Scalp Avenue, Johnstown, PA 15904, phone (814) 269–2545, fax (814) 269–2798, e-mail **theochem@ctc.com**. Peter S. Shenkin, Department of Chemistry, Barnard College, New York, NY 10027, phone (212) 854–1418, e-mail **shenkin@still3.chem.columbia.edu**

To Subscribe: Send the message `SUBSCRIBE MMODINFO` to the address **MMODINFO-REQUEST@UOFT02.UTOLEDO.EDU**

To Unsubscribe: Send the message `UNSUBSCRIBE MMODINFO` to the address **MMODINFO-REQUEST@UOFT02.UTOLEDO.EDU**

To Post: Send your message to **MMODINFO@UOFT02.UTOLEDO.EDU**

Comments: Unmoderated. Archived.

Archives: Anonymous FTP **FTP.UTOLEDO.EDU**

Name: MOL-DIVERSITY—Molecular Diversity List

Description: To facilitate free exchange of ideas among molecular diversity researchers all over the world.

Subjects: Random peptide and protein libraries; oligosaccharide and oligonucleotide libraries; small organic molecular libraries; library design, construction, and automation; pooling and deconvolution strategies; molecular tag, scaffolding, and building blocks; high-throughput and computational screening; library database and software management tools; announcement of new conferences and job openings
Editor: Kit S. Lam, phone (602) 626–2333; Zhan G. Zhao, e-mail **Zhao@biosci.arizona.edu**
To Subscribe: Send the message: `SUBSCRIBE MOL-DIVERSITY Firstname Lastname` to **Listserv@listserv.arizona.edu**
To Unsubscribe: Send the message: `SIGNOFF MOL-DIVERSITY` to **Listserv@listserv.arizona.edu**
To Post: send messages to **MOL-DIVERSITY@listserv.arizona.edu**
Comments: Unmoderated.

Name: MOLECULAR-DYNAMICS-NEWS—News in Molecular Dynamics
Description: News and discussion for researchers involved in the study of the dynamics of molecules, mostly in the gas phase. Subjects covered include molecular collisions, reaction dynamics, high-resolution spectroscopy of Van der Waals complexes, and molecular photodissociation. The list is used mostly for conference announcements and details of positions vacant at the faculty and postdoctoral levels.
Subjects: chemical physics, molecular collisions, photodissociation, Van der Waals complexes
Editor: Jeremy M. Hutson, Department. of Chemistry, University of Durham, Durham, DH1 3LE, UK, e-mail **.M.Hutson@durham.ac.uk**
To Subscribe: Send the message `join molecular-dynamics-news Firstname Lastname` to the address **mailbase@mailbase.ac.uk**
To Unsubscribe: Send the message `leave molecular-dynamics-news` to the address **mailbase@mailbase.ac.uk**
To Post: Instructions are sent when you join.
Comments: Unmoderated. No subscription restrictions. 10 messages/month. 700 subscribers. Archived.
Archives: Archived automatically. Access via e-mail (`MAILBASE GET` command; send the message `send mailbase user-guide` to **mailbase@mailbase.ac.uk** for instructions on e-mail retrieval), anonymous FTP **mailbase.ac.uk**, gopher **mailbase.ac.uk**, WWW **http://mailbase.ac.uk/**

Name: MXDIAG-L—Molecular Pathology and Diagnostics
Description: For discussion of molecular diagnostics, including scientific, technical, and regulatory issues.
Subjects: DNA, RNA, analysis

Comments: Unmoderated. 5 messages/month. 150 subscribers. Archived. List type: LISTSERV.
Administrative address: **LISTSERV@ALBNYDH2.bitnet**
Distribution address: **MXDIAG-L@ALBNYDH2.bitnet**
Archives: To retrieve messages for a given month, send the message `GET MXDIAG-L LOGyymm` (where yy is year and mm is month) to **listserv@albnydh2.bitnet**, for example, `GET MXDIAG-L LOG9409`

Name: NAPRONET—Natural Products
Description: An electronic scientific forum to discuss chemistry and biology of natural products. It is intended to be used for discussions about chemistry and biological activity of natural products, academic and industrial professional issues, contact searches, international and national meeting announcements, and for various information exchanges within the scope of the list.
Subjects: chemistry of natural products, biological activities of natural products, spectroscopy, isolation techniques, syntheses of natural products, chemical transformations of natural products, biological transformations, biotechnology
Editor: Mahmut Miski, Sanofi Winthrop Pharmaceuticals, R&D, 25 Great Valley Pkwy., Malvern, PA 19355, phone (610) 889–6101, fax (610) 889–6009, e-mail **mmiski@mcimail.com** or **miski@netcom.com**.
Mustafa Akgul, Department of Industrial Engineering, Bilkent University, 06533 Bilkent, Ankara, Turkey, phone 90–312–266 4000 ex1237, fax 90–312–266 4477, Secretary 90–312–266 4126,
e-mail **Akgul@bilkent.edu.tr**
Comments: Unmoderated. 25 messages/month. 80 subscribers. Archived. List type: LISTSERV.
Administrative address: **listproc@bilkent.edu.tr** or **listserv@trbilun.BITNET**
Distribution address: **NAPRONET@bilkent.edu.tr** or **NAPRONET@trbilun.bitnet**
Archives: Send the message `INDEX NAPRONET` to **listproc@bilkent.edu.tr** or **listserv@trbilun.BITNET** and retrieve available file with the `GET NAPRONET filename` command, for example, `GET NAPRONET 940827` will retrieve messages from Aug. 27, 1994.

Name: Neuron—Digest of Neural Networks News
Description: A list (in digest form) dealing with all aspects of neural networks (and any type of network or neuromorphic system).
Subjects: natural systems, software simulations, neurobiology, hardware, neuroscience, digital, artificial systems, analog, neural networks, optical, algorithms, cellular automatons, Hebbian systems, Widrow–Hoff algo-

rithm, perceptron, threshold logic, holography, content-addressable memories, Lyapunov stability criterion, Navier–Stokes equation, annealing, spin glasses, locally coupled systems, globally coupled systems, dynamic systems, (adaptive) control theory, back-propagation, generalized delta rule, pattern recognition, vision systems, parallel distributed processing, connectionism

Editor: Peter Marvit, Psychology Department, University of Pennsylvania, 3815 Walnut Street, Philadelphia, PA 19104, phone (215) 573–3991, fax (215) 898–7301, e-mail **neuron-request@psych.upenn.edu**

To Subscribe: Send a short message to **neuron-request@psych.upenn.edu**

To Unsubscribe: Send a short message to **neuron-request@psych.upenn.edu**

To Post: Send your message to **neuron-request@psych.upenn.edu**

Comments: Moderated. 4–8 digests/month. 2400 direct subscribers (but list has many redistribution points and readership may be as high as 40,000). Archived.

Archives: Anonymous FTP **psych.upenn.edu**, [130.91.68.31] in the directory `pub/Neuron-Digest`. That directory contains back issues with the names vol-nn-no-mm (e.g., vol-3-no-02). I'm also collecting simulation software in `pub/Neuron-Software`. Contributions are welcome. Back issues can also be obtained through the mail archive server. Send a message to **archive-server@psych.upenn.edu** with `help` as the subject line for more information.

Name: ORGCHEM—Organic Chemistry

Description: To experiment with electronic exchange of information related to organic chemistry concerns of all types. The orgchem list also has an anonymous FTP site associated with it to facilitate interchange of graphical materials such as *ChemDraw* documents or other graphical formats. The ultimate goal of this list is to provide an electronic version of dropping into a neighbor's office for advice and insight.

Subjects: organic chemistry, physical organic chemistry

Editor: Curt Breneman, Department of Chemistry, Cogswell Lab 319A, Rensselaer Polytechnic Institute, Troy, NY 12180, phone (518) 276–2678, fax (518) 276–4045, e-mail **breneman@xray.chem.rpi.edu** or **brenec@rpi.edu**

To Subscribe: Send a message to **orgreq@extreme.chem.rpi.edu** requesting membership

To Unsubscribe: Send a message to **orgreq@extreme.chem.rpi.edu** requesting deletion

To Post: Send a message to **orgchem@extreme.chem.rpi.edu**

Comments: Unmoderated. No subscription restrictions. 30 messages/month. 200 subscribers. Archived.

Archives: orgchem messages are saved and graphical files are archived in appropriately named subdirectories of the anonymous FTP server at **extreme.chem.rpi.edu**

Name: ORG-GEOCHEM—Organic Geochemistry

Description: An electronic discussion list for organic geochemists and other scientists with interests in organic geochemistry. Related fields such as environmental geochemistry and petroleum geochemistry are covered. Org-Geochem uses the Mailbase system to manage message Distribution. Most messages relate to conference details, research opportunities, and publication information. Some technical messages.

Subjects: organic geochemistry, environmental geochemistry, petroleum geochemistry

Editor: Bryn Jones, Fossil Fuels and Environmental Geochemistry, The Drummond Building, The University, Newcastle upon Tyne, NE1 7RU, UK, phone 091 222 6000 ex 8628, fax 091 261 1182, e-mail **Bryn.Jones@newcastle.ac.uk**

To Post: Send messages by e-mail to**org-geochem@mailbase.ac.uk**

Comments: Unmoderated. Up to 10 messages/month. 200 subscribers. Archived. List type: MAILBASE.
Administrative address: **mailbase@mailbase.ac.uk**
Distribution address: **org-geochem@mailbase.ac.uk**

Archives: Archived automatically. Access via e-mail (`MAILBASE GET` command—send the message `send mailbase user-guide` to **mailbase@mailbase.ac.uk** for instructions on e-mail retrieval), anonymous FTP **mailbase.ac.uk**, gopher **mailbase.ac.uk**, WWW **http://mailbase.ac.uk/**

Name: PHARMACY MAIL EXCHANGE—Pharmacy Discussions

Description: A general discussion forum for pharmacists and workers in related fields.

Subjects: pharmacy teaching and practice, pharmacology, pharmaceutical chemistry

Editor: Paul Hodgkinson, Department of Pharmacy, De Montfort University, The Gateway, Leicester LE1 9BH, UK, phone +44 533 577285, fax +44 533 577287, e-mail **phh@dmu.ac.uk**. TWX: 9312132271 PH G

To Subscribe: Send message with details to **pharm-request@dmu.ac.uk**

To Unsubscribe: Send message to **pharm-request@dmu.ac.uk**

To Post: Details are provided after subscription.

Comments: List restricted to professionals and students. 20 messages/month. List messages sent also to sci.med.pharmacy newsgroup

Archives: gopher **gopher.dmu.ac.uk**, anonymous FTP **ftp.dmu.ac.uk**, WWW **http://www.dmu.ac.uk/0/departments/pharmacy/archive/www/pharm.html**

Name: PMD—Parallel Molecular Dynamics

Description: Announcements and discussion about the parallel molecular dynamics program PMD. PMD is an experimental program written in C and implementing the fast multipole algorithm of Greengard and Rokhlin. It is easily portable between different UNIX system architectures and message-passing protocols. PMD calculates a good approximation of the full long-range Coulomb interactions faster than conventional, much less accurate, cutoff methods. PMD is also scalably parallel with distributed data storage. It has been run on workstation networks as well as dedicated parallel machines such as the CM-5 and the Intel Paragon. PMD is made freely available to all interested researchers to provide a flexible testing ground for advanced algorithms.

Subjects: PMD (parallel molecular dynamics)

Editor: Andreas Windemut, Columbia University, Department of Biochemistry and Biophysics, 630 West 168th Street, BB-221, New York, NY 10032, phone (212) 305–6884, fax (212) 305–6926, e-mail **windemut@cumbnd.bioc.columbia.edu**

To Subscribe: Send a message to **pmd-request@cumbnd.bioc.columbia.edu**

To Unsubscribe: Send a message to **pmd-request@cumbnd.bioc.columbia.edu**

To Post: Send mail to **pmd@cumbnd.bioc.columbia.edu**. Posting is discouraged at this point. Better to send mail to the coordinator at the address **pmd-request@cumbnd.bioc.columbia.edu** and announcements of general interest will be made.

Comments: Unmoderated. No subscription restrictions. One message/month. 110 Subscribers. No archives yet.

Name: POLYMERP—Polymer Physics

Description: A discussion group on polymer physics. The topics include meetings, articles, software, theories, materials, methods, tools, and polymer properties such as solubility, diffusion, viscosity, and adsorption. Application programs and data of general interest ("critical tables") can be sent to the list owner for archiving. Files are listed in PolymerP filelist. Volunteers willing to maintain a file are encouraged to contact the list owner.

Subjects: polymers, materials, diffusion, viscosity

Editor: Peter Barneveld, e-mail **Peter@Fenk.WAU.NL**

Comments: Unmoderated. No subscription restrictions. 350 subscribers. Archived. List type: LISTSERV.
Administrative address: **LISTSERV@HEARN.NIC.SURFNET.NL**
Distribution address: **POLYMERP@HEARN.NIC.SURFNET.NL**

Archives: Send the message `INDEX POLYMERP` to the address **LIST-SERV@HEARN.NIC.SURFNET.NL** to obtain a list of available files. Then send the message `GET POLYMERP filename` to the address **LISTSERV@HEARN.NIC.SURFNET.NL** to retrieve individual files.

Name: POST—Postdoc International

Description: A job announcement service for postdoctoral fellows. The jobs are organized by country and category, for example, usa.chemistry, gb.chemistry, and france.biology. To retrieve the file containing job announcements, send a message to **post@docserv.saclay.cea.fr**. For example, send the message

```
get usa.chemistry canada.chemistry usa.computers
```

if you want to find out about jobs in chemistry in Canada, chemistry in the United States, and computer-related jobs in the United States.

Editors: Claire Gautherin **GAUTHERI@AMOCO.SACLAY.CEA.FR**, Dominique Marchand **DOMI@AMOCO.SACLAY.CEA.FR**, Rosa Antidormi **ROSA@AMOCO.SACLAY.CEA.FR**, Vincent Vigneron **VVIGNE@AMOCO.SACLAY.CEA.FR**, Etienne Amic **AMIC@AMOCO.SACLAY.CEA.FR**

To Post: The list is not intended for discussions. To announce a new position, start your message with the word `message` on a line by itself and place your announcement on the line below.

Comments: Very useful service for those who need jobs or those who need them filled fast.

Name: PROCESS-L—Process control

Description: To encourage and facilitate the discussion of information regarding all aspects of process control and technology, including on-, in, or at-line sampling and analysis technologies; process design, integration, and modeling; and control theory and practice.

Subjects: process analysis, process control, process technology, IPC, process integration, process sampling, industrial chemistry, industrial control

Editor: Andy Packham, Department of Instrumentation and Analytical Science, University of Manchester, Institute of Science and Technology, Manchester, PO Box 88, M60 1QD, UK, phone 061 200 4899, fax 061 200 4911, e-mail **mcnajp@fs4.in.umist.ac.uk**

Comments: Unmoderated. 10 messages/month. 100 subscribers. Not archived but some files available by anonymous FTP **fs4.in.umist.ac.uk**.
List type: MAISER.
Administrative address: **MAISER@FS4.IN.UMIST.AC.UK**
Distribution address: **Process-L@FS4.IN.UMIST.AC.UK**

Name: QUANTA-L—Discussions of Quanta Users

Description: For users of QUANTA and related programs (e.g., ChemNote). Discussion on CHARMm is also welcome here. The MSI Scientific Support Bulletin Board mailing lists are intended to provide an informal forum for users of MSI products to discuss various aspects of the programs. The list is not designed to be an official channel for product support; all requests for technical product support should continue to go through the Scientific Support hotline.

Subjects: molecular mechanics, molecular dynamics, proteins, nucleic acids

Editor: Robert B. Funchess, Molecular Simulations Inc., 16 New England Executive Park, Burlington, MA 01803–5297, phone (617) 229–9800 ex 202, fax (617) 229–9899, e-mail **bobf@msi.com**, WWW **http://www.msi.com/~bobf/bobf.html**

Comments: Unmoderated, 10 messages/month, 200 subscribers, Archived. List type: LISTPROC.
Administrative address: **listproc@msi.com**
Distribution address: **quanta-l@msi.com**

Archives: Archived monthly. Anonymous FTP **server.msi.com**, WWW **http://www.msi.com/**, or via e-mail to **listproc@msi.com** (send `index` and then request needed files by sending the `get filename` command to **listproc@msi.com**).

Name: RADCH-L—Radiochemistry

Description: Discusses problems of radiochemistry.

Subjects: radiochemistry

Editor: **ABBE@FRCPN11.BITNET** or **ABBE@frcpn11.in2p3.fr**

Comments: Unmoderated. 120 subscribers. List type: LISTSERV.
Administrative address: **LISTSERV@frcpn11.in2p3.fr**
Distribution address: **RADCH-L@frcpn11.in2p3.fr**

Name: SAFETY—Laboratory Safety

Description: Discussion of environmental health and safety issues, particularly as they relate to academic and research settings. Emphases include laboratory safety, hazardous waste disposal, and regulatory compliance. A file containing a description of the list is available as `SAFETY WELCOME` from **LISTSERV@UVMVM.UVM.EDU**

Subjects: laboratory safety, regulatory compliance

Editor: Ralph Stuart, Environmental Safety Facility, 655 Spear Street, University of Vermont, Burlington, VT 05405, phone (802) 656–5403, fax (802) 656–5407, e-mail **rstuart@moose.uvm.edu**

Comments: Unmoderated. No subscription restrictions. 500 messages/month. 1300 subscribers. Archived. List type: LISTSERV.

Administrative address: **LISTSERV@UVMVM.UVM.EDU**
Distribution address: **SAFETY@UVMVM.UVM.EDU**

Archives: Send the message `INDEX SAFETY` to **LISTSERV@UVMVM.UVM.EDU** to get file list, and then retrieve files by sending `GET SAFETY filename` to **LISTSERV@UVMVM.UVM.EDU**.
There is also the gopher **SIRI.UVM.EDU**

Name: SUP-COND—Superconductivity in Israel

Description: A mailing list established in Israel on Bitnet for people interested in superconductivity so that both local and international information on this field (local achievements, abstracts on work from overseas, lectures, workshop, calls for proposals, etc.) can be disseminated efficiently and rapidly.

Subjects: superconductivity

Editor: Joseph van Zwaren, Coordinator for Physics Research, National Council for Research and Development, Ministry for Science and Development, e-mail **JO@ILNCRD.BITNET**

To Post: Send any newsworthy item to **SUP-COND@TAUNIVM**

Comments: Moderated. No subscription restrictions. 250 subscribers. No archives. List type: LISTSERV.
Administrative address: **LISTSERV@VM.TAU.AC.IL**
Distribution address: **SUP-COND@VM.TAU.AC.IL**

Name: SYBYL—Discussions of Sybyl Users

Description: A forum for exchange of information pertaining to use of the Sybyl molecular modeling package from Tripos Associates. This list is not an official means of communication with Tripos, but members of the technical and sales staff do answer questions posted to the list.

Subjects: molecular modeling, medicinal chemistry, molecular mechanics, molecular dynamics, Sybyl, Tripos, drug design

Editor: Curt Breneman, Department of Chemistry, Cogswell Lab 319A, Rensselaer Polytechnic Institute, Troy, NY 12180, phone (518) 276–2678, fax (518) 276–4045, e-mail **breneman@xray.chem.rpi.edu** or **brenec@rpi.edu**

To Subscribe: Send a message to **sybylreq@extreme.chem.rpi.edu** requesting membership

To Unsubscribe: Send a message to **sybylreq@extreme.chem.rpi.edu** requesting deletion

To Post: Send your message to **sybyl@extreme.chem.rpi.edu**

Comments: Unmoderated. No restrictions on subscribing. 50 messages/month. 250 subscribers. Archived.

Archives: Sybyl messages and contributed SPL scripts are saved in subdirectories of the anonymous FTP server at **extreme.chem.rpi.edu.**

Name: WATOC—World Association of Theoretical Organic Chemists
Description: Mailing list for the World Association of Theoretical Organic Chemists
Subjects: computational chemistry, organic chemistry, inorganic chemistry, theoretical chemistry
Editor: Henry Rzepa, Department. Chemistry, Imperial College, LONDON SW7 2AY, UK, phone (44) 171 594 5774 or 594 5809, fax (44) 171 594 5804, e-mail **rzepa@ic.ac.uk**.
WWW **http://www.ch.ic.ac.uk/rzepa.html**
To Subscribe: Send the message `subscribe watoc Firstname Lastname` to the address **listserver@ic.ac.uk**
To Unsubscribe: Send the message `unsubscribe watoc` to the address **listserver@ic.ac.uk**
To Post: Send the message to **watoc@ic.ac.uk**
Comments: Moderated. No subscription restrictions. Low traffic. 400 subscribers.
Archives: Available on WWW **http://www.ch.ic.ac.uk/watoc.html**

Name: WMCIC-L—Women in Chemistry in Canada
Description: A communication link for women in chemistry to tell each other about job opportunities in their universities, news about upcoming speakers or events of interest, etc. Particularly beneficial to graduate students looking for postdoctoral positions and professors looking for new graduate students and postdoctoral fellows.
Subjects: job opportunities in chemistry for women
Editor: Angela Lyon, e-mail **lyon@chem.queensu.ca**
Comments: Unmoderated. No subscription restrictions. 2 messages/month. 25 subscribers. Archived, but archives are not accessible publicly.
List type: LISTSERV.
Administrative address **Listserv@qucdn.queensu.ca**
Distribution address: **wmcic-l@qucdn.queensu.ca**

Name: WSN—Water Science Network
Description: To facilitate communication between experimental and theoretical scientists with different backgrounds, whose research interests are related to water structure and hydrogen-bond network, thermodynamic and kinetic properties of water, phase transitions in water, hydrophobic effects, hydration forces, water around solutes and water at interfaces, hydration of biomolecules, water in membranes, water in cells, experimental techniques for studying water properties, computer simulations of water. Postings contain discussions of new ideas and developments in related fields, questions and answers about particular problems, biblio-

graphic information on current and forthcoming books and articles, information on funding sources, meeting announcements, position announcements, and product and service announcements.

Subjects: water science, that is, chemical physics, molecular physics, computational chemistry, physical chemistry, structural chemistry, cell biology, molecular biology, etc.

Editor: Iosif Vaisman, Laboratory for Molecular Modeling, School of Pharmacy, University of North Carolina at Chapel Hill, Chapel Hill, NC 27599-7360, phone (919) 966–7821, fax (919) 966–6919, e-mail **wsn-adm@mmlds1.pha.unc.edu** or **vaisman@gibbs.unc.edu**

To Subscribe: Send the message `subscribe water Firstname Lastname` to the address **listserv@unc.edu**

To Unsubscribe: Send the message `unsubscribe water` to the address **listserv@unc.edu**

To Post: Send message to **water@listserv.unc.edu**. Only subscribers may post.

Comments: No restrictions on subscribing, 480 subscribers, 15–20 messages/month.

Archives: Available from LISTSERV (send a message with the single line `index water` to **listserv@unc.edu** and then use the `GET` command. FTP, gopher, and WWW access under development.

Name: XPLOR-L—Discussions of X-PLOR users

Description: For users of X-PLOR (stand-alone or using the X-ray structure or NMR structure interfaces between QUANTA and X-PLOR). This is an MSI Scientific Support Bulletin Board mailing list intended to provide an informal forum for users of MSI products to discuss various aspects of the programs. The list is not designed to be an official channel for product support; all requests for technical product support should continue to go through the Scientific Support hotline.

Subjects: molecular mechanics, molecular dynamics, proteins, nucleic acids

Editor: Robert B. Funchess, Molecular Simulations Inc., 16 New England Executive Park, Burlington, MA 01803–5297, phone (617) 229–9800 ex 202, fax (617) 229–9899, e-mail **bobf@msi.com**, WWW **http://www.msi.com/~bobf/bobf.html**

Comments: Unmoderated, 10 messages/month, 200 subscribers, Archived.
List type: LISTPROC.
Administrative address: **listproc@msi.com**
Distribution address: **xplor-l@msi.com**

Archives: Archived monthly. Anonymous FTP **server.msi.com**, WWW **http://www.msi.com/**, e-mail **listproc@msi.com** (send `index` and then request needed files by the sending `get filename` command to **listproc@msi.com**).

Name: YSN—Young Scientists' Network

Description: (1) To let the press, public, and government officials know that there is currently no shortage of scientists. Young scientists are currently encountering unprecedented difficulty finding jobs. (2) To discuss how young scientists can find both traditional and nontraditional careers. The primary activity of the YSN is the production of the electronic Young Scientists' Digest, which is sent out five times per week. The Moderated YSN (MYSN) Digest, a weekly summary of the regular YSN Digest, is edited by Jennifer Cohen. A job list, which contains job openings that have been submitted to the YSN, is maintained by Mary Ellen Scott. The YSN archive is maintained by Arthur Smith.

Subjects: employment issues, social issues, alternative careers, miscellaneous

Editors: John Quackenbush, Stanford Human Genome Center, Stanford University Department of Genetics, 855 California Ave., Palo Alto, CA 94304, phone (415) 812–1938, fax (415) 812–1916, e-mail **ysnadm@crow-t-robot.stanford.edu** or **johnq@crow-t-robot.stanford.edu** for private messages. MYSN Moderator: Jennifer M. Cohen, 22 S. Prince Street, Shippensburg, PA 17257, phone (717) 530–5098, e-mail **jenc@pinet.aip.org**. Joblist Administrator: Mary Ellen Scott, University of Akron Chemistry Department., Akron, OH 44325–3601, phone (216) 972–8392, e-mail **mes@atlas.chemistry.uakron.edu** Archive administration: Arthur P. Smith, Chemistry Department, BG-10, University of Washington, Seattle, WA 98195

To Subscribe: Send a message with the `Subject:` line `subscribe Firstname Lastname YSN` if you want to get full discussions, or with the `Subject: subscribe Firstname Lastname MYSN` if you want the weekly digest to the address **ysnadm@crow-t-robot.stanford.edu**

To Unsubscribe: Send a message with the `Subject:` line: `unsubscribe YSN` or `unsubscribe YSNM` to **ysnadm@crow-t-robot.stanford.edu**

To Post: Send your message to **ysn@crow-t-robot.stanford.edu**

Comments: Unmoderated (YSN) and moderated (MYSN). 22 messages/month on YSN. 8 messages/month MYSN. 2500 subscribers on YSN and 1750 on MYSN. Archived.

Archives: Anonymous FTP **snorri.chem.washington.edu**, WWW **http://snorri.chem.washington.edu**

Acknowledgments

The core of this chemistry listing was supplied by the list editors and maintainers. We are very grateful for all who responded to our query.

We ask that you help in keeping this compilation current by sending announcements and changes to J. K. Labanowski by e-mail at **jkl@osc.edu**.

chapter ten

Managing the Computational Chemistry List

Jan K. Labanowski, Charles F. Bender, and Alejandro Pisanty

In This Chapter

This chapter describes the experience of managing an electronic list from the point of view of the list administrator, Jan K. Labanowski.

3223–7/96/0235$12.00/1

THE COMPUTATIONAL CHEMISTRY LIST (CCL) is an electronic discussion forum in which electronic mail (e-mail) messages sent to the address **chemistry@ccl.osc.edu** are distributed among the subscribers. The administrative details are taken care of at **chemistry-request@ccl.osc.edu**.

About the List Itself

The list has an associated set of archives that contain all past messages, a substantial selection of public domain and shareware software for computational chemists, documentation, coordinates of interesting compounds, miscellaneous data files, text-processing tools, information pointers, a list of conferences in the field of computational chemistry, a list of job openings and jobs sought, notes on available software, and so on. The archives are accessible by standard network protocols such as file-transfer protocol (FTP), Gopher, World Wide Web, and electronic mail. The archives can also be searched by keywords and regular expressions.

The list was established in January 1991 as an off-hours project by the administrator with the hardware and network access support of the Ohio Supercomputer Center. The stable environment of the supercomputer center, the broad scientific and technical in-house experience, and network and computer resources were important factors in the list's success. Since that time, the list has been steadily growing. It is difficult to estimate the exact number of participants, because many addresses that are subscribed to the list are local lists or local newsgroups. Local lists and newsgroups subscribe primarily to save bandwidth, as only one copy of the message needs be sent to a particular site. As of October 1994, more than 2000 addresses from more than 40 countries subscribed to the list. About half of the addresses belong to domains located within the United States. About 250 addresses are in U.S. commercial enterprises (the **.com** domain), 750 in academia (the **.edu** domain), 70 in government organizations (the **.gov** domain), 20 in military organizations (the **.mil** domain), and 30 in other nonprofit organizations (the **.org** domain).

Turnover is relatively small. People who change addresses tend to resubscribe at their new addresses. The traffic on the list varies depending on the day of the week. Lately, CCL has been receiving about 10–20 messages per day, and that number has been rising because of the steady increase in the number of list subscribers.

The list is operated by the software developed in-house written in Perl, although security-sensitive portions of the software are written in C. The list is not moderated, and the sender addresses are not verified against the list of subscribers, because doing so would prevent participants who monitor the list via local reflectors, Gopher, World Wide Web, or newsgroups from sending messages to the list.

However, the list is monitored. Before a message is distributed among subscribers, many automated tests are done to weed out undesirable messages (e.g., empty body messages, subscription messages, wrongly addressed bounces, messages larger than 100 kbytes, and messages without a subject line). The software is also capable of blocking messages from certain addresses. The rules of the list and instructions are available in the list's `help` file, which is sent to each new subscriber. This file is also available from the list archives. The list is not currently mirrored as a general access Usenet group. Mirroring, that is, allowing unrestricted posting to the list from a publicly accessible Usenet newsgroup, would expose the list to more random messages, advertising, chain letters, and other inappropriate material. However, any nonsubscriber can view the list messages with the help of a Gopher or World Wide Web browser because the messages are archived automatically as they arrive.

Dealing with Traffic on the List

To limit the list traffic to matters of general interest, the administrator (J. K. Labanowski) offers his help in finding addresses and contacts and in answering standard questions concerning the topic of the list. Also, the experience and software developed for running this list is being shared. Other lists have adopted the distribution and maintenance software developed for running the CCL, and some lists have used the rules of the CCL as a blueprint for their own set of rules.

The traffic on the list is steadily growing, and it may force some subscribers to leave the list. Some of this traffic is not related to the list topic. Those who accidentally or deliberately violate the rules are reminded by the list administrator that their messages are inappropriate. The offenders are also blocked from direct posting to the list, and their messages then need to be reviewed by the administrator (when time permits) before they are distributed. This problem would be solved by moderating the list. However, besides being a time-consuming task for a high-volume list, it would also deprive the electronic list of its interactivity. The list administrator is obviously not available 24 hours a day while the list is operating continuously in all time zones. If only one reviewer were to moderate the list, the list could be biased toward that person's interests. In this case, review would not be anonymous, so it could create difficult situations for the reviewer and the submitter.

It is also not clear, from a legal point of view, who bears the legal responsibility when sensitive information is sent to the list (e.g., copyrighted software, trade secrets, and controversial material). Inappropriate messages can be weeded from the list in many ways (e.g., by employing anonymous reviewers selected automatically from the list subscribers in time zones coinciding with working hours), but this topic is controversial, and the software to perform these tasks has yet to be written.

Another way of dealing with high volume is to create sublists. In principle, it is possible to use software that could assign a message to the proper sublist (or sublists, if the message overlaps several fields) by analyzing the text of the message. Subscribers could subscribe to one or more sublists and receive messages that are within their reading profiles. Such software could also refuse messages that cannot be categorized. The rules for assigning a message to a given category or topic would have to be developed.

A survey of the list readership conducted during the summer of 1994 (discussed in the next section) is a good starting point for identifying possible subtopics for the list. Obviously a simple keyword search is not likely to produce satisfactory results. Software needs to be written to perform context-sensitive text analysis. Another easier solution is to request that authors properly tag their messages. However, forcing subscribers to adhere to rules is difficult, especially if the rules are complex (e.g., the author would have to review the list of current subtopics before posting).

Obviously, the list maintainer has a different sense of what the list should be than the participants have. Tools are available for searching list archives, but subscribers prefer to ask the same questions over and over again rather than search the archives for previous discussions on a topic. Also, many trivial questions are finding their way to the list. The solution to this problem, that is, to provide an extensive electronic tutorial, would require substantial effort. The flavor of unconstrained discussion is an important advantage of the list, but the growing volume of discussions may discourage people with busy schedules from participating. Reminders occasionally posted by the list maintainer may improve the situation temporarily, but this tool cannot be used too frequently, for it would defeat the list purpose.

Survey of the CCL's Impact on Chemical Research and Teaching

In the summer of 1994, we ran a survey of the CCL's impact on chemical research and teaching. We sent a set of questions by electronic mail to all list subscribers, of whom approximately 15% replied. The questions explored the ways in which the Internet is changing research and teaching. We intend here to show the main lessons from the survey that are of interest to people providing Internet services. We present only a short summary here, and the details of the survey results and discussion can be found in CCL archives as **ftp://ccl.osc.edu/pub/chemistry/documents/CCL-survey** and via the CCL home page: **http://ccl.osc.edu/chemistry.html**.

The survey responses are representative of the geographic distribution of the subscribers, which is 50% in the United States, followed by 10% each in Germany, the United Kingdom, and Canada. More than half of the respondents work in academic institutions, and close to 20% work in for-profit organizations. The results of the survey should therefore be extrapolated carefully to other services if they are aimed to a more commercial kind of readership.

The CCL has commanded a large loyalty from its readers. Two-thirds of the respondents have been on the list for more than 1 year. The respondents hold a wide variety of positions, mostly at the research scientist and graduate student levels. Although the list has computational chemistry as its main subject, less than a third of its participants consider computational chemistry their field of activity. This fact shows that computational chemistry has spawned many subfields, and its tools are used by many who are not computational chemists.

Most of the respondents would be characterized as "power users" in many organizations outside academia, because the equipment and software they use are mostly UNIX "boxes". However, many actually access the list through lower-level equipment, for example, personal computers at home connected to the Internet through modems. Most of the respondents read their electronic mail daily, many of them even several times a day.

The levels of usage of the list were classified according to the number of messages actually posted by the respondents and the number of messages to which they reply. Our classification leads to the following types: Normal, Responder, Lurker, Talkative, Shy, and Extremely Light. Most of the respondents to the survey are of the Normal type (send or reply to five or fewer postings in a period of about a year), and 15% classify themselves as Lurkers, which means they only read the list and do not write or post to it. Of course, the number of Lurkers is far larger; they also benefit significantly from the list.

The CCL has become an important, almost vital, resource to its users. It is, in their words, "a class in computational chemistry where each person gets a turn at teaching" and "the commons of the computer-savvy chemical crowd". It is considered useful because it combines the opinions of experts on fundamental matters, as well as the advice of the people most handy with the technical issues that arise during the actual conduct of research.

One prominent participant of the CCL made an explicit comparison of the way in which he used the CCL to solve a problem within 1 hour and the traditional approach without the CCL, which needed a full week of work. However, not only speed but also the quality of the aid is generally acknowledged.

The other important specific benefits from using the list are finding or acquiring software and data and directions to literature from prominent experts. Furthermore, the CCL has become a medium for contacts leading to collaborations between people who geographically are far removed from each other. In this last sense, the CCL has also been valuable for people who are in distant locations or otherwise isolated (e.g., being the only computer-savvy chemist in an organization) to keep in touch with advances in the field and to stay within the computer-chemical community.

The list has benefited from an active self-regulation, by which irrelevant postings, repeated or irrelevant discussions, flaming and other disagreeable conduct, and inadequate commercial usage have become marginal.

People who would like to emulate the success of the CCL in providing similar services in other fields could start by the most important factor: relevant postings by relevant people, in two tiers: fundamental (almost philosophical) questions of the field, and practical, technical issues that demand immediate answers (coping with bugs in programs and the like), with a reasonable but strict netiquette that keeps the list readable and worthy of attention all the time. This goal requires a high-quality intervention from the list coordinator or moderator, a good archive system, and the recruiting of the best possible list members from the start.

Administrative and Technical Requirements

There is a great potential for the list to become a continuous scientific conference. It may also become a shared archive for publicly available chemical software, data, and standards. It may become a meeting place for prospective scientific collaborations, a high-quality news service, and an educational forum. However, all this cannot be achieved as an off-hours project by the list maintainer. These activities would require a substantial administrative, research, and technical effort. Priority items include the following:

The list must be moderated in real time by anonymous reviewers. For the list to have an interactive feel, the acceptance of the article will have to be done within a few hours, a not few days, from the moment of submission.

- Sublists must be created for several different topics, or tools must be provided for list subscribers to pre-sort the messages to their particular areas of interest. There are many possibilities in this area, but the appropriate protocols and software need to be developed.
- List distribution must be decentralized and hierarchical. It does not make sense to send the same message over transcontinental lines a few hundred times. National or regional distribution centers need to be negotiated for the list. These centers should be capable of assuming the leadership if needed. This arrangement would make the list less dependent on the stability of any particular person or organization. We are not suggesting that the current personnel or organization may change, but it is always better to have a vision for the distant future.
- List archives need to be mirrored on different continents and countries; thus access time will be faster. With the expected growth of Internet traffic this issue may become important.

- An answer desk must be created where beginners can ask questions. This facility would ensure high professional standards for the list without compromising its educational role.
- New list-administration software must be developed. The options that are currently available do not offer an integrated solution to the administration of a list of this size.
- A dedicated archivist must be hired to maintain the archives. The archives are an important aspect of the list. They contain thousands of valuable tips, answers, and data, but they also contain material of lesser quality. Active effort is needed to cross-reference, index, and interface the archives. Making them easily accessible to users requires maintenance and software development.

Efforts are under way to find appropriate funding for the list. This funding will be important during the transition period. Once the software is developed and the list is distributed geographically, the effort will become manageable for local institutions participating in the computational chemistry network. The computational chemistry information network can also be subsidized by the sale of archives on CD–ROM and nonintrusive advertising via the archives. With the projected progress in network access and speed, and the growing importance of computational chemistry in basic research and industrial research and development, the need for professional maintenance of the list becomes significant. Only time will show if the list can sustain its growth and still be a tractable resource for its subscribers.

chapter eleven

Chemistry and Gopher

Steven M. Bachrach

In This Chapter

This chapter gives a list of gopher sites that contain chemistry-related information and presents a discussion of some of them.

3223–7/96/0243$12.00/1

MANY GOPHER SITES are devoted to chemistry. Most of these are run by chemistry departments at academic institutions. The most frustrating aspect of the chemistry GopherSpace (the network of all gopher servers) is the lack of interesting and useful information contained on these sites. Nevertheless, a number of interesting gopher sites are operational. I will discuss a few of these and give a larger list, including addresses and ports, in the Appendix. The Internet is constantly in flux, and therefore, gopher servers may come and go, addresses may change, and some of the information in this appendix may no longer be valid.

American Chemical Society Gopher. Run by the American Chemical Society (ACS), this gopher delivers information pertaining to the ACS publications and member services. ACS books can be searched using a Wide Area Information Service (WAIS) system. Order forms for books and journals can be downloaded. Supplementary material for the *Journal of the American Chemical Society* is available in the `ACS Publication/Supplementary Materials` directory. Instructions to authors for the ACS journals can be viewed.

Computational Chemistry List Archive. The organizers of the Computational Chemistry List (discussed in Chapter 10) have collected materials associated with this list and made it available by Gopher. The archives of all of the messages from the list are indexed and can be searched and retrieved. Tables of contents for *Theoretica Chemica Acta*, the *Journal of Computational Chemistry*, and the *International Journal of Quantum Chemistry* can be obtained. This server also operates a postdoctoral clearinghouse in computational chemistry.

First Online Chemistry Conference on Chemical Education. In the summer of 1993, the first electronic chemistry conference was held (discussed in Chapter 8), making use of FTP and electronic mail (e-mail) for distributing the papers and discussions. The archives of this conference are available by Gopher.

Genethon Gopher. This site contains a listing of tables of contents for many journals, especially those related to biochemistry. Most listings are not current, except for the *Journal of Biological Chemistry*, which is maintained up-to-date.

Indiana University Chemistry Library Gopher. This gopher site contains many resources pertaining to traditional library resources, including guides to the chemistry literature; how to use Gmelin, Beilstein, and *Chemical Abstracts*; and lists of books that help with chemical nomenclature and locating chemical suppliers.

***Journal of Chemical Education* Gopher.** The *Journal of Chemical Education* operates a gopher site that contains information related to the journal. This information includes instructions to authors, tables of contents, and subscription information. The site also contains information on *JCE Software* and maintains a forum on topics related to chemical education.

***Journal of Chemical Physics* Gopher.** The *Journal of Chemical Physics* operates a project called *JCP Express* whereby manuscripts that have been accepted for publication in *JCP* but have not yet appeared in print are made available through their gopher. Instructions for authors and tables of contents are also available.

University of Missouri–St. Louis Chemistry Gopher. This gopher contains an ongoing effort to compile a comprehensive list of all chemistry textbooks in print.

National Science Foundation. The NSF gopher includes program guidelines and a searchable index of its publications and awards.

Northern Illinois University Chemistry Gopher. This gopher server is part of a pilot project to explore the use of the Internet as a means of disseminating chemical information. To that end, a number of resources have been established at this site. These include a chemists' e-mail phone book, an academic chemical employment clearinghouse, a conference listing, journal submission guidelines, chemical stock prices, and a database of chemical software.

QCPE Gopher. The Quantum Chemistry Program Exchange (QCPE) is an archive and distribution source for a wide variety of computational chemistry software. Its catalog and ordering instructions are available at this gopher site.

University of Utah Chemistry Department. The outstanding feature of this gopher site is the Materials Safety Data Sheet (MSDS) archive. This archive contains unlicensed and unrestricted MSDSs for a large number of chemicals. Everyone is encouraged to contribute to this archive by sending in MSDSs in ASCII format.

Appendix: Chemistry-Related Gopher Sites

Name	*Address*	*Port*
American Chemical Society	acsinfo.acs.org	70
Australian Defense Force Academy Chemistry Department	gopher.adfa.oz.adfa.oz.au (1/LibraryServices/Electronic Library/subject/sci/chem)	70
Brookhaven Protein Data Bank	irisc2.chm.bnl.gov	70
Case Western University Chemistry Department	chemgopher.chem.cwru.edu	70
Computational Chemistry List Archive	infomeister.osc.edu	73
Chemistry at CSC	gopher.csc.fi (1/tiede/kemia)	70
First Online Chemistry Conference on Chemical Education	info.umd.edu (1/inforM/Educational_Resources/ Faculty_Resources_and_Support/ ChemConference)	901
Genethon Gopher	gopher.genethon.fr	70
Genome Data Bank Gopher	gopher.gdb.org	70
Imperial College Chemistry Gopher	155.198.1.33	70
Indiana University Chemistry Library Gopher	lib-gopher.lib.indiana.edu	7050
Institut für Physicalische und Theoretische Chemie (Regensburg) Gopher	rchs1.chemie.uni-regensburg.de	70
Internet Network Information Center	is.internic.net (1/infoguide/resources/chemistry)	70
Journal of Chemical Education Gopher	jchemed.chem.wisc.edu	70
Journal of Chemical Physics Gopher	jcp.uchicago.edu	70
Michigan State University Chemistry Department	slater.cem.msu.edu	70

Name	*Address*	*Port*
National Science Foundation	stis.nsf.gov	70
Northern Illinois University Chemistry Gopher	hackberry.chem.niu.edu	70
Oklahoma State University Chemistry Department	jgelder.chem.okstate.edu	70
QCPE Gopher	lib-gopher.lib.indiana. edu (1/qcpe)	7050
Rutgers University Chemistry Gopher	ndb.rutgers.edu	70
University of California–Davis Chemistry Department	gopher-chem.ucdavis.edu	70
University of California–Los Angeles Chemistry Department	gopher.chem.ucla.edu	70
University of California–Santa Cruz Chemistry Department	papilio.ucsc.edu	70
University of Missouri–St. Louis Chemistry Department	gopher://slvaxa.umsl.edu/ 11gopher_root_chem%3a%5 bbooks%5d	70
University of Oklahoma Chemistry Department	cheminfo.uoknor.edu	70
University of Utah Chemistry Department	atlas.chem.utah.edu	70
University of Wisconsin–Madison Chemistry Department	gopher.chem.wisc.edu	70
Yale University Chemistry Department	yaleinfo.yale.edu (1/Chemistry)	70

chapter twelve

Chemistry and the World Wide Web

Henry S. Rzepa

In This Chapter

Some of the novel chemical applications developed during the pioneering early years of 1993–1994 are explored, including concepts such as hyperactive molecules, living science journals and books, hypermedia-based education resources, on-line conferences, and the use of databases and hyperglossaries.

3223–7/96/0249$13.50/1

THERE IS LITTLE POINT in competing with the World Wide Web (WWW, the Web) system itself in trying to present an up-to-date snapshot of how it is being applied to chemistry. In this chapter, therefore, a short review of some unusual and innovative applications to chemistry is presented to introduce some of the implications of this technology.

The uniform resource locators (URLs) for some chemistry and related WWW sites are given in Table 12.1. A number of similar collections of chemical information exist on the Web itself. For example, the Yahoo collection (accessed with the URL **http://www.yahoo.com/Science/Chemistry**; *see* Table 12.1) illustrates the diversity already apparent in the subject (Figure 12.1). Currently the most comprehensive list of pointers is at the University of California at Los Angeles, with more than 300 educational sites and some 50 links to commercial organizations and a smaller number of nonprofit organizations (as of September 1995).[1]

WWW Browsers and Chemical Collaboratories

The first generations of WWW browsers essentially involved a single user interacting with a distributed source of information, but as noted in Chapter 6, they did not readily allow for any form of collaborative browsing between multiple users, or of more complex interactions between a browser and its associated helper programs. The OpenDoc style compound document architectures may mature by 1996, but an interesting interim solution has recently been suggested. Known as the Common Client Interface, or CCI,[2] it is currently supported only by the National Center for Supercomputing Applications (NCSA) `Mosaic` browser. In its early implementations, the browser would be configured by the user to accept or to broadcast CCI instructions on a specified port number. Current implementations include a slide-show program that can choreograph a sequence of pages in the form of successive URLs on a `Mosaic` browser and an application that allows multiple student sessions of `Mosaic` to copy the actions of a teacher session. Because there is no explicit need for the teacher and student systems to be physically adjacent, this technology opens up many interesting teaching possibilities.

The first chemically specific program to support the CCI interface was the `Eyechem` suite,[3] which allows hyperlinks defined with the body of a protein databank (pdb) format file to be expanded for visual-

Table 12.1. URLs for Chemistry and Related Subject Sites

Site	*Description*
http://www.ch.ic.ac.uk/WWW/	Update to this chapter
http://www.w3.org/hypertext/DataSources/bySubject/Overview.html	Overall subject overview
http://www.ncsa.uiuc.edu/SDG/Software/Mosaic/Docs/whats-new.html	Global What's New pages
http://akebono.stanford.edu/yahoo/new.html	Yahoo What's New pages
http://www.chem.ucla.edu/chempointers.html	American collection of chemistry pointers
http://www.che.ufl.edu/WWW-CHE/index.html	American collection of chemical engineering pointers
http://specter.dcrt.nih.gov:8004/	American collection of molecular biology pointers
http://www.yahoo.com/Science/Chemistry/	Yahoo collection of chemistry pointers
http://www.shef.ac.uk/uni/academic/A-C/chem/chemistry-www-sites.html	European collection of chemistry pointers
http://apamac.ch.adfa.oz.au/	Australian collection of chemistry pointers
http://www.acs.org/	American Chemical Society
http://chemistry.rsc.org/rsc/	Royal Society of Chemistry
http://hackberry.chem.niu.edu/Infobahn/Paper38	"Best of the Web" Poster, ACS Meeting, 1995
http://www.lycos.com/	Lycos Global Search Engine specific information sources
http://www.chem.ucla.edu/chem_news.html	Chemistry-related newsgroups
http://www.electrochem.org/ecs/	Electrochemical Society
http://euch6f.chem.emory.edu/hetsoc.html	International Society of Heterocyclic Chemists
http://info.cas.org/welcome.html	Chemical Abstracts Service Home Page
http://osc.edu/chemistry.html	Computational Chemistry Archive
http://www.sura.net/matsci.html	Materials Science Research Database
http://www.spo.eds.com/patent.html	Electronic Data Systems Shadow Patent Office
http://galaxy.einet.net/galaxy/Science/Chemistry.html	EINet Chemistry Home Page
http://hackberry.chem.niu.edu:70/0/webpage.html	Chemistry at NIU
http://www-jbc.stanford.edu/jbc/	*Journal of Biological Chemistry*
http://www.awod.com/netsci	Network Science
http://chemistry.uca.edu/mjce/	*Multimedia Journal of Chemical Education*
http://www.csc.fi/lul/chem/graphics.html	Chemistry Art Gallery
http://www.nih.gov/molecular_modeling/mmhome.html	Molecular Modeling at NIH
http://www.arc.umn.edu/GVL/Software/xmol/XMol.html	`XMol` on-line manuals
http://www.emsl.pnl.gov:2080/	Gaussian basis sets
http://expasy.hcuge.ch/pub/Graphics/IMAGES/GIF	SWISS-IMAGE
http://www.ch.ic.ac.uk/ectoc/	ECTOC organic chemistry conference

continued on next page

Table 12.1. continued

Site	*Description*
http://bellatrix.pcl.ox.ac.uk/people/barry_spring95/barry.gig_econf.html	EGC glycoscience conference
http://www.bio.cam.ac.uk/scop	Structural markup of proteins
http://www.unige.ch/crystal/crystal_index.html	Crystallography Home Page
http://csdvx2.ccdc.cam.ac.uk/	Cambridge Crystallographic Data Centre
http://ibc.wustl.edu/klotho/	Biochemical Compounds Database
http://www.arc.ab.ca/~morgan/Nano.html	Nanotechnology Archives
http://www.hgmp.mrc.ac.uk/	Human Genome Mapping Project
http://www.silverplatter.com/sampler/webspirs.html	Silver Platter Database
http://www.ch.ic.ac.uk/chemical_mime.html	Hyperactive Molecules
http://www.wired.com/vrml and http://www.sdsc.edu/vrml/	Virtual reality modeling language
http://www-hpcc.astro.washington.edu/scied/science.html	Useful index of science-teaching materials
http://www.chem.vt.edu/chem-ed/vt-chem-ed.html	Virginia Tech education resources
http://www.ch.ic.ac.uk/GIC/	Instructional chemistry at Imperial College
http://www.cchem.berkeley.edu/Education/index.html	Resources for education at Berkeley
http://www-wilson.ucsd.edu/education/samplegateway.html	UCSD MultiMedia Chemistry Lab
http://www.utexas.edu/world/instruction/	World Lecture Hall
http://www.automatrix.com/conferences/	Collection of conference information
http://www.ch.ic.ac.uk/talks/	Chemistry talks and posters
http://golgi.harvard.edu/biopages.html	Biosciences
http://cui_www.unige.ch/meta-index.html	Global Index Center

ization using either `Eyechem` or `Mosaic` as appropriate. Alternatives to the CCI formalism have also been proposed, and this is currently an area under active development.

Living Journals, Active Conferences, Real-Time Workshops, and Global Talks

During the ≈200-year development period of chemistry as a subject, natural divisions have developed between the three primary mechanisms for disseminating information.

Primary research journals and to a lesser extent books still form the basic mechanism by which new results are subjected to peer review

Figure 12.1. The Yahoo collection of chemical pointers.

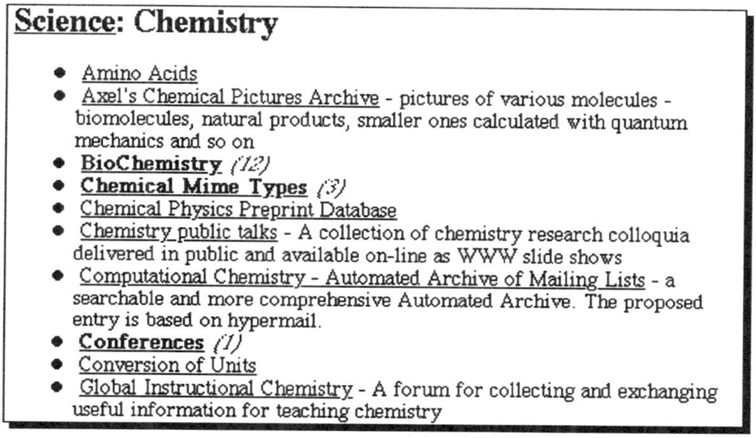
Science: Chemistry

- Amino Acids
- Axel's Chemical Pictures Archive - pictures of various molecules - biomolecules, natural products, smaller ones calculated with quantum mechanics and so on
- **BioChemistry** (12)
- **Chemical Mime Types** (3)
- Chemical Physics Preprint Database
- Chemistry public talks - A collection of chemistry research colloquia delivered in public and available on-line as WWW slide shows
- Computational Chemistry - Automated Archive of Mailing Lists - a searchable and more comprehensive Automated Archive. The proposed entry is based on hypermail.
- **Conferences** (1)
- Conversion of Units
- Global Instructional Chemistry - A forum for collecting and exchanging useful information for teaching chemistry

and are formally recorded in print. However, even with all the advances in publishing that have occurred over the past few decades, it can still take up to a year for the printed form of scientific journals or books to become available. Supplementary data associated with the original research may take longer to appear, probably in a form that is difficult to access, or it may never be published. Color continues to be expensive, and authors frequently have to pay charges for the inclusion of color plates. Perhaps the most serious limitation of the printed page is that numerical and symbolic information is vulnerable to transcription errors when retrieved from the page for local use.

Conferences and workshops offer a more visual and interactive form of communication between a physically present audience, but the printed proceedings can suffer the same publication delays as journals, and the cost of attending a major international congress can be substantial. The use of "parallel sessions" in many conferences means that instant decisions regarding the choice of material have to be made by delegates on the basis of often little information.

Invited research colloquia or time-tabled teaching lectures to small audiences at an institute may never reach a wider audience, and notes taking during such events might be subject to error. Nevertheless, feedback and discussion during such sessions remains a valuable asset that should be preserved.

The Web has the potential to play a major unifying role in this regard. In 1994, a number of scientific journals began to offer contents lists, preprints,[4] and other information in on-line form, albeit via the anonymous FTP (file-transfer protocol) or Gopher mechanism rather than using HTTP–HTML. Starting in January 1994, a number of primary research publications were marked up in HTML and mounted on Web servers as pilot experiments.[5] These papers had previously been refereed in the conventional manner and were cross-referenced to each other via hyperlinks. By providing a URL with the submission, the actual refereeing of the paper was accomplished by the anonymous referees connecting and browsing through the hyperlinks. In another paper,[6] the coordinates associated with ab initio calculations were made available in the form of hyperactive molecules (Figure 12.2) via the chemical MIME (multipurpose Internet mail extension) protocol.

Figure 12.2. A hyperactive molecule hyperlinked to an electronic journal article.

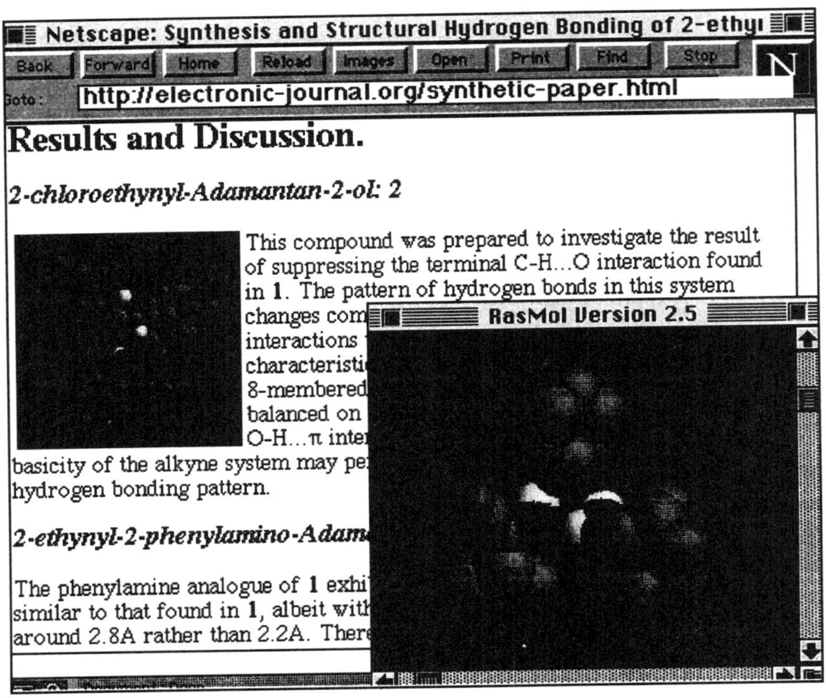

Protein Science[7] was the first chemically oriented journal scheduled to appear in 1995 in HTML marked-up form, with the protein structures available via the symbolic Kinemage[8] notation developed for the journal, and delivered to the user by means of hyperlinks to the file. Currently the best example of this genre is the *Journal of Biological Chemistry* (Figure 12.3). The home page contains a wealth of information, including effective keyword search areas, a "guest book" that visitors can sign, information on the necessary browser configuration, and highly accessible indicators of the content of the journal. This journal represents a translation of an existing printed journal to the new medium, whereas *Network Sci-*

Figure 12.3. The home page for an electronic journal.

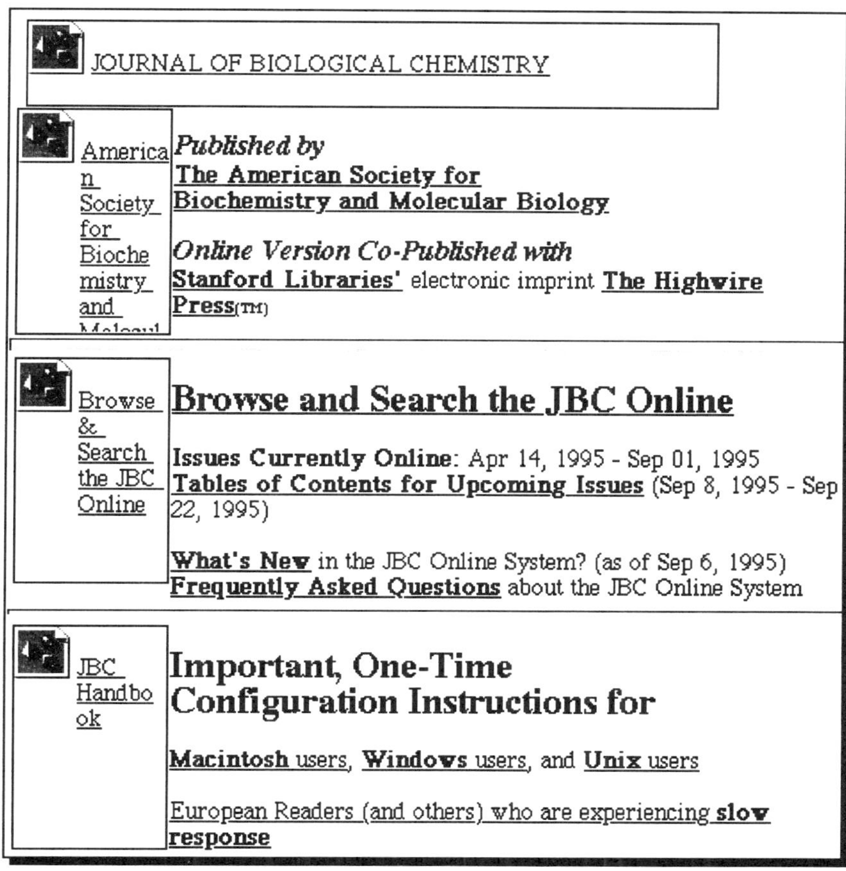

ence represents an innovative current awareness forum with no printed equivalent, for which the production costs, at least currently, are obtained by advertising rather than recovery from the readership. The *Multimedia Journal of Chemical Education* (Table 12.1) is another attempt to move away from the printed page, offering authors an opportunity to illustrate original teaching techniques with actual working examples of their materials. Other international chemical journals were also expected to appear in on-line form during 1995, offering condensed graphical abstracts of papers,[9] diagrammatic and visual material that is otherwise expensive to print, or extensive markup of the entire article. The large number of limitations associated with HTML 2.0 is only partially addressed with HTML 3.0, and the full maturity of electronic journals may not be accomplished for several years.

Other important problems must be solved if such mechanisms are to be used more extensively. Perhaps the most significant of these is cost recovery for the publishers. During 1995, commercial HTTP servers offering document encryption and data-stamping became available, and it is also possible to restrict access to documents by validating the IP (Internet protocol) address and name of the computer making the originating request. Document encryption enables individual financial transactions, but IP address validating would allow licensing on a departmental or an institutional scale.

Also not to be underestimated is user resistance to a purely electronic viewing mechanism, and most likely printed versions of journals will be around for some time yet, although probably running in parallel with an on-line version. Nevertheless, traditionally expensive components of paper publishing such as color plates and distribution, along with the newer media of audio and animation, are particularly well served by mechanisms such as the WWW. Current implementations of HTML (mostly at version 2.0 of the definition) do not really allow a comfortable "look-and-feel" to be imparted to the formatted page, and various experiments with alternative implementations of SGML (standard generalized markup language) markup such as the ISO 12083 standard[10] for scholarly publishing are underway. These have hitherto required proprietary browsers, and few such solutions offer the Internet access given by the Web or have support for chemical MIME.

Electronic journals are likely to continue to represent the formal refereed aspect of scientific publishing, and more informal newsletter-style publications also seem set to flourish, of which a good example is the home page of the International Society of Heterocyclic

Chemists.[11] In particular, issues such as cost recovery and perhaps even permanence are less important with this form of publishing.

The Web mechanism is also being actively promoted for the organization of conferences. During 1994, the registration, programs, and agendas for two international conferences on the WWW itself (at CERN in Geneva, Switzerland, in May[12] and in Chicago, IL, in October,[13]) were made available entirely on-line, although the events themselves occurred at physical locations and on specific dates, and the proceedings were eventually published on paper in a conventional journal.[14] The chemistry workshop at the first of these conferences allowed contributions to be discussed and edited, at a considerable saving in travel costs for the contributors.[15] Conference papers, posters, and electronic-mail discussion for the First Electronic Computational Chemistry Conference (ECCC)[16] were accomplished using the Web mechanism alone, although in this instance the proceedings were subsequently published in CD-ROM rather than printed form, and the electronic version was then withdrawn.[17] This conference introduced for the first time a powerful method of integrating list-server-based electronic-mail discussions by conversion to HTML documents using a program called `Hypermail`.[18]

Another model of the use of the Web for conferences was the "Infobahn" session at an ACS national meeting.[19] This session consisted of live presentations to an audience by speakers combined with "electronic" posters that both the audience and outside "visitors" could browse. Such an integration of live and electronic components seems likely to form a permanent feature of most conferences in the future.

The ECCC conference was followed by one focused on organic chemistry,[20] where further innovations such as full indexing, molecular hyperglossaries, and the enclosure of chemical files with electronic-mail contributions were introduced. A typical range of features possible within such conferences is shown in Figure 12.4. The on-line nature of such conferences presents unusual problems for organizations such as Chemical Abstracts Service, who act as curators of the world's chemical information. The current policy is to abstract such conferences if the material is scheduled for "hard" publication in printed or CD-ROM form, but it seems likely that some form of abstracting will also eventually need to evolve for more evanescent information available exclusively via the Internet.

Live presentations to audiences using the networked WWW as a slide-show presentation package can be accomplished by means of a

Figure 12.4. The home page for a virtual chemistry conference.

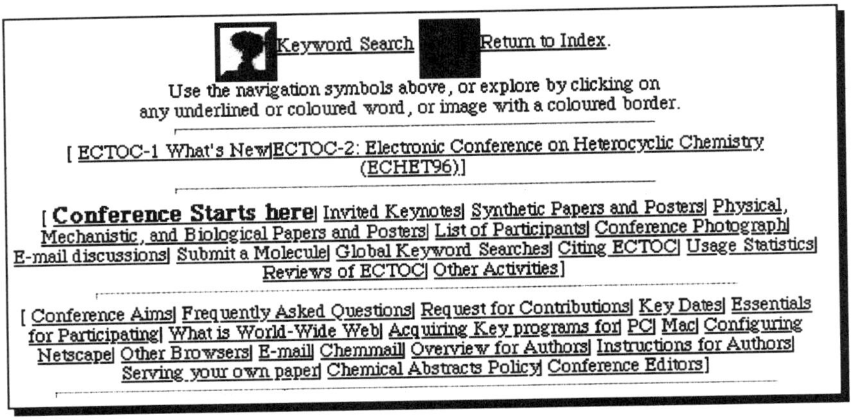

computer with overhead projection facilities.[21] The audience is given an opportunity to explore the detail of the talk at their leisure, and the lecturer has an opportunity to concentrate on discussion with the audience rather than delivering the detail. It may also be significant that one of the first applications of the CCI interface[2] was the production of a WWW slide-show organizer. Software products such as `PageMill`[22] may point the way that authoring environments for slide-based material might evolve.

No absolute dividing lines can be drawn in the continuum represented by scholarly electronic journals at one end and networked presentations at the other. Perhaps ultimately, the issue of quality alone will determine how these mechanisms are used and viewed. After many generations when scientific authors and their publishers enjoyed a mutually symbiotic relationship, we are also entering a new era in which the interests of these two groups may be showing signs of diverging. The rules are rapidly being redefined, and the dust may settle in surprising ways.

Chemistry Art Galleries and Visual Displays

The visual impact of chemistry is substantial, and a number of sites offer guided tours through a collection of stunning images. Justice cannot be done to all the sites within the limited space of this chapter, but several WWW sites in particular set the trend and deserve a mention.

The SWISS-IMAGE[23] collection of annotated protein diagrams is an impressive indication of how effectively the WWW medium can serve to arouse interest in a subject. These images are largely in bit-mapped format, but one site has specialized in producing animated images corresponding to dynamic chemical phenomena.[24]

At another site, [25] a comprehensive and well-presented on-line guide to the representation of molecular models and rendering techniques allows colorful illustrations of the various examples to accompany the text (Figure 12.5).

The Image Library of Biological Macromolecules is a good example of a growing three-dimensional macromolecule database,[26] and an excellent collection of hydrogen-like atomic orbitals, including f-functions, is available at Virginia Tech.[27] These atomic orbitals are available not only as two-dimensional images, but in the new VRML format that allows the reader to inspect the functions in three-dimensional format, truly a three-dimensional sculpture in the chemical art gallery.

Figure 12.5. Index page to Rendering Molecular Models.

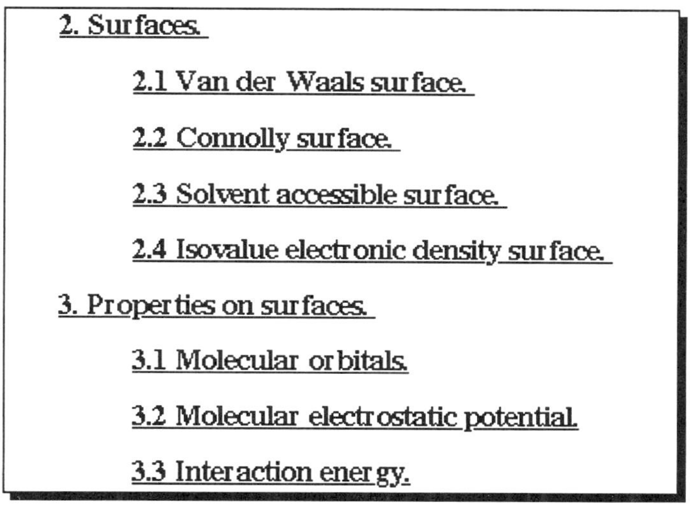
2. Surfaces.

2.1 Van der Waals surface.

2.2 Connolly surface.

2.3 Solvent accessible surface.

2.4 Isovalue electronic density surface.

3. Properties on surfaces.

3.1 Molecular orbitals.

3.2 Molecular electrostatic potential.

3.3 Interaction energy.

Chemical Databases and Manuals

A natural application of the WWW is the provision of database material and documentation in on-line format. It is becoming common for program documentation and manuals to be converted to HTML format and made available via the WWW. For example, the `XMol` and `RasMol` modeling programs have on-line manuals available in this form. An on-line searchable directory of registered chemists is available, and newcomers can register via a Forms interface.[28]

Some excellent examples of chemical databases appeared during 1994, including an innovative user interface to the protein database at the National Institutes of Health[29] termed "Molecules R Us". The WWW home page for this service offers a simple a keyword index search term, which expands into a list of items that match the search term presented in hyperlinked form. If one of these is selected, the user is then given the opportunity of either downloading the pdb (protein databank) coordinate file in chemical MIME format as a hyperactive molecule for local visualization and manipulation, or of selecting a viewing angle and having a graphics interchange format (GIF)-type graphic image generated for downloading (Figure 12.6). Such creation of documents-on-demand from a remote database

Figure 12.6. Graphical Interface to a pdb database.

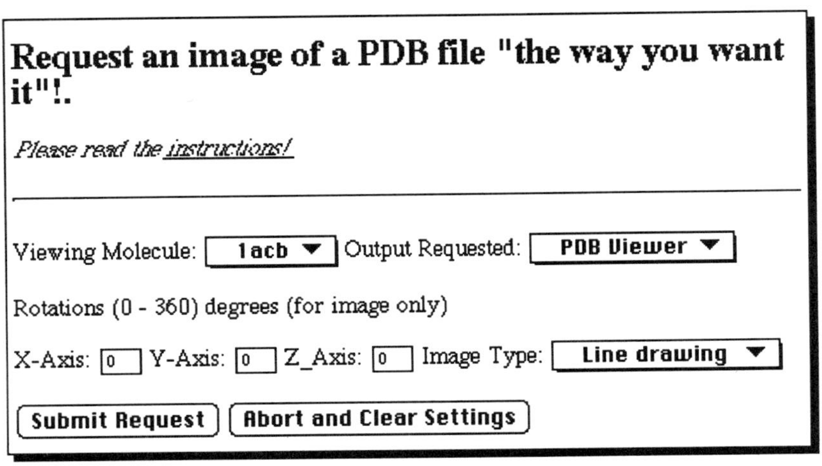
Request an image of a PDB file "the way you want it"!.

Please read the instructions!

Viewing Molecule: 1acb Output Requested: PDB Viewer

Rotations (0 - 360) degrees (for image only)

X-Axis: 0 Y-Axis: 0 Z_Axis: 0 Image Type: Line drawing

Submit Request | Abort and Clear Settings

greatly simplifies the maintenance of information and reduces the propagation of errors.

The concept of taking molecular structures and adding value by suitable annotation and representation is also illustrated by the SCOP (structural classification of proteins) project, the aim of which is to provide a detailed description of the structural and evolutionary relationships between proteins whose structure is known.[30] A similar format has been provided for access to Gaussian basis functions for use in ab initio calculations, via a forms interface in which the user selects the basis they are interested in (Figure 12.7).[31]The CORINA facility[32–34] is another service by which user-supplied two-dimensional molecular connectivity data can be transformed into reliable three-dimensional coordinates.

Another pointer to the future is the biochemical compounds declarative database known as Klotho.[35] Here automatic generation of a SMILES string and a Fischer projection of compounds retrieved from a search is achieved using a layered transformational grammar. (SMILES is simplified molecular line entry specification.) The SMILES strings are passed to *CONCORD*, which generates a pseudo-energy-minimized structure, and the resulting pdb-type files can then be passed to a visualizer such as *RasMol* using chemical MIME types.

Figure 12.7. Graphical interface to Gaussian basis set library.

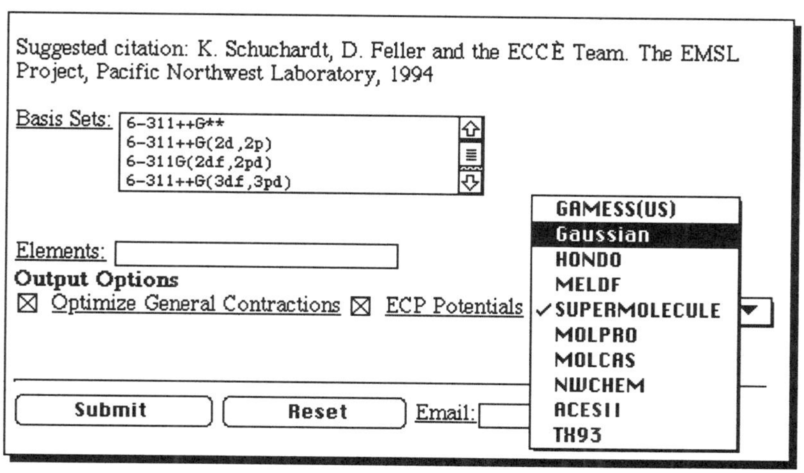

Some innovative experiments in offering commercial databases are also available. Daylight Information Systems has produced a Web-based "user interface" to several of their databases. Information retrieved for individual compounds is converted on-the-fly to two-dimensional image representations, along with generated SMILES strings and other attributes.[36] A similar interface has been introduced by SilverPlatter to their own range of bibliographic information.[37]

One advantage of using the Web is that such sources of information can be linked into local documentation to provide a consistent and integrated local information environment. The "electronic chemistry library" is an example of a student laboratory exercise that links together a number of sources of information into the context of a university laboratory preparation of a penicillin derivative.[38] Such a "shopping mall" metaphor for acquiring chemical information may form the basis for future database environments.

Teaching Applications

Chemistry faculties in both the United States and Europe (Table 12.1) are using WWW as a shop window to highly visual and often interactive chemistry multimedia projects, in which the enthusiastic contributions of students can play a major role (Figure 12.8). The WebElements Periodic Table[39,40] project at Sheffield University illustrates the quality of information that is now achievable. This project comprises a 30-megabyte collection of 5500 files presented as a deceptively simply periodic table of the elements and containing extensive hyperlinks to facilitate the retrieval of information.

Such materials can also be consolidated into a CD-ROM-based reference work. All the hyperlinks can continue to be active, and some can point to updates resident on the network rather than the CD-ROM medium.[41] Such a combination of local and network-based material may become increasingly necessary as loading of the Internet results in unacceptable response times for exclusive network transfer of large images, animation, and other materials. Such collections of materials on the Internet are often referred to as "virtual libraries", and as conventional libraries, books, and journals become ever more expensive to maintain, it is tempting to conclude that the virtual model has a number of irresistible attributes, of which cost is a major factor. A more imponderable feature is whether either the permanency or the

Figure 12.8. A selection of globally available teaching materials.

11. Laboratory and Lecture Courses at the University of West Indies, Jamaica.
12. WebElements - a periodic table and isotope database
13. Welcome to the World of Physical Chemistry at UCSD
    - Quantum Mechanics
    - X-Ray Diffraction
    - Molecular Dynamics
    - Statistical Mechanics
    - Spectroscopy
14. A complete course on the Principles of Protein Structure
15. A Glossary of Biologically important Molecules and Terms and index search
16. World Lecture Hall in Chemistry

quality of virtual materials compares at all well with the conventional sources of information.

A rather different model is illustrated by the GlobeWide Network Academy,[42] which offers a remote teaching paradigm of distributed teaching. Course material is placed on-line, and students can acquire it as part of their studies using WWW browsers. An interesting innovation is that the material can be prepared by instructors at widely different locations and at different times. The material can be discussed either by using more conventional e-mail mechanisms, or via so-called MOOs (multiuser object-oriented environments),[43] which involve the student joining live on-line discussions using text as the communication medium. This concept is perfectly exemplified by the Virtual School of Natural Sciences' Principles of Protein Structure (VSNS-PPS) course.[44] Some 250 registered students from all over the world are joined by 50 course "consultants", and a considerable quantity of original and high-quality material is available for all participants. Coursework is set at regular intervals and assessed, and an interactive "hyperglossary" of terms used in the course materials can be contributed to by all.

As both the quality and the breadth of such pedagogic content increases, it will increasingly become possible to design new teaching or laboratory courses by suitably editing Web-resident materials. Most likely, a proportion of such materials will have restricted access, and it may be necessary to purchase some components of courses in the future. Many might assume that campus-centered teaching is unlikely to be greatly threatened by such developments in the short term, but significant funding and resource is beginning to flow into this area. Some people already believe that much teaching practice will change dramatically in response to these technologies.

Molecular Hyperactivity

Representing the intrinsically three-dimensional nature of most molecules has never been satisfactory on the printed page. One method of overcoming this problem would be to create an MPEG (motion picture experts group) video animation of a molecule or a derived property, which would impart a three-dimensional perception to the image. Such a mechanism allows the precise viewing angle of any individual frame to be determined by the original author. This capability has distinct advantages when that viewing angle is of critical importance to the particular scientific point being made. On other occasions, the user may wish to explore the three-dimensional characteristics of molecules or multidimensional potential energy surface.[45] Such navigability is enabled by the chemical MIME and the VRML mechanisms noted earlier in this chapter. In the following sections, some of the features of this new medium are explored.

Conformational and Structural Information

Molecular hyperactivity[46] is defined as a WWW hyperlink to active molecular information, implemented in such a way that users can acquire and analyze the information themselves. The example in Figure 12.9 shows how a set of hyperlinks is mapped into a two-dimensional conformational energy surface; each hyperlink is associated with a set of molecular coordinates. When the hyperlink is activated, the coordinates are transferred locally, and a suitable viewing program is activated. This program enables the reader to explore the geometrical features of the potential energy surface (Figure 12.9). More importantly, the essential numerical integrity of the original report can be checked if required by passing the coordinates to appropriate computational programs, in this case *Mopac*. The local user can rapidly add further value via other types of computation not reported by the original authors.

These steps can all be accomplished with an immediacy that contrasts with the procedure associated with scientific printed journals.

Figure 12.9. A conformational energy surface containing hyperlinks to molecular coordinates.

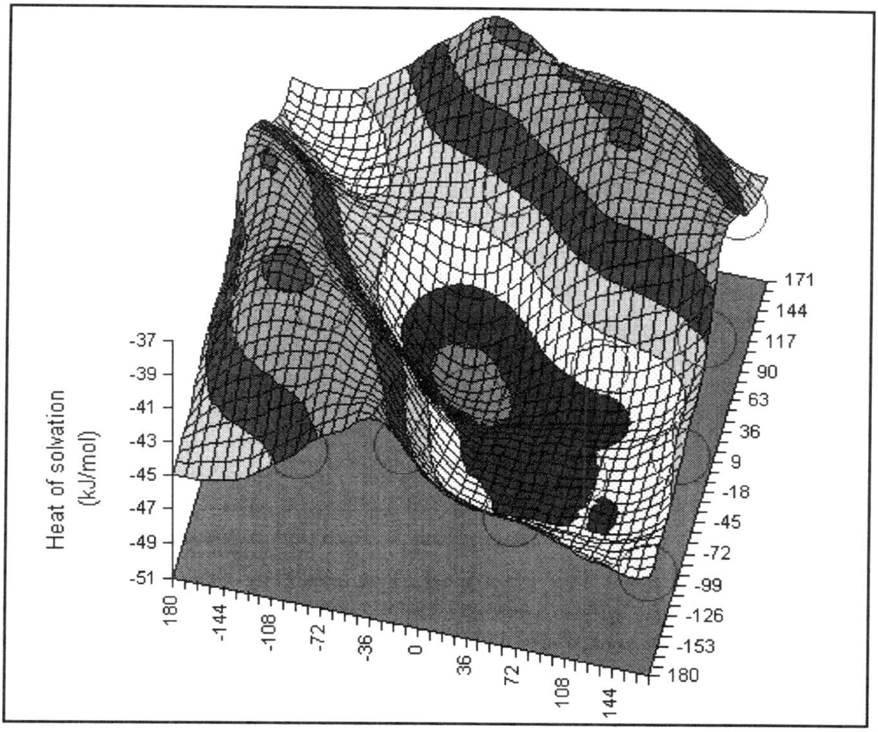

With printed journals, authors of the original paper must take the trouble to deposit molecular coordinates as supplemental information with the journal, conventionally on paper. The interested reader would then have to order this information, scan the contents into a computer file or type the numerical data manually, correct any errors, and finally process it using a suitable program. Such a process could easily take 2 or more years from the original date of the calculations.

In this example, the molecular coordinates were acquired from calculations, but they could equally well derive from crystallographic or other spectroscopic studies. A substantial proportion of printed papers report results derived from structural features of molecules. Often, novel features are not noticed by the original authors and can go undiscovered for several years after the original report. The association of "hyperactive molecules" with electronic articles may go some way to enabling more effective and timely use of structural information.

NMR Spectroscopy

The two-dimensional mapping technique used in the previous example can be applied to the analysis of spectroscopic information and is particularly well-suited to two-dimensional NMR analysis (Figure 12.10). In this implementation, the hyperactive molecule barnase is first activated using the molecular helper `RasMol`[47] by clicking on the thumbnail in the bottom right of the image. Individual regions of the two-dimensional NOESY (nuclear Overhauser effect spectroscopy) correlated ^1H spectrum corresponding to cross-peaks are mapped via hyperlinks to a set of files stored on the server, in which the relevant portion of the protein is identified. These files contain markup instructions to the `RasMol` program, and hence formally constitute a chemical structure markup language, or CSML.[46] Currently, `RasMol` does not support HTTP and hence cannot receive

Figure 12.10. A two-dimensional ^1H NOESY NMR spectrum hyperlinked to structural assignments.

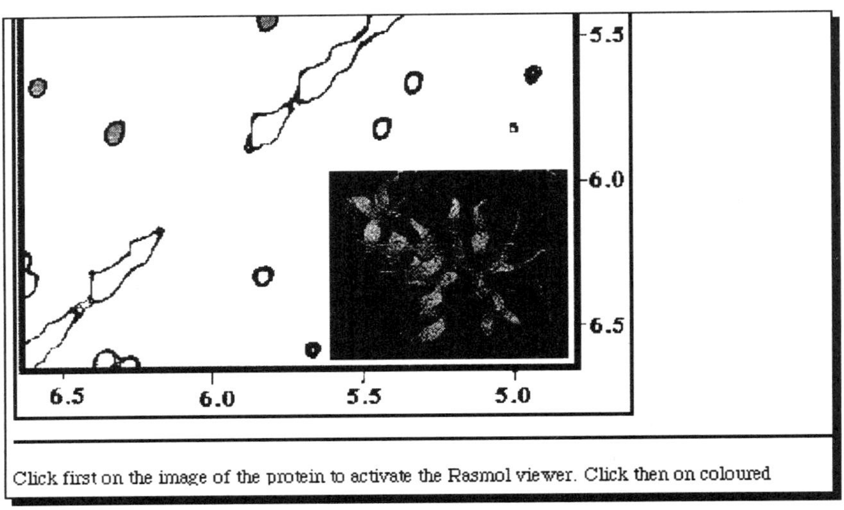

this markup directly. Instead, the markup files are assigned a chemical MIME type

```
chemical/x-csml
```

and transferred to the local computer using the HTTP mechanism present in the WWW browser. The browser is configured to pass this particular MIME type to a small script on the user's machine, which in turn passes the chemical markup instructions to the `RasMol` process. Such an indirect mechanism is also necessary to preserve user security. There is a pressing need to define a generic chemical markup language as an adjunct to the VRML language[48] that is already in use. One such initiative is already underway.[49]

A rather different approach is exemplified in the JAVA project[50], which defines a language that can be compiled and downloaded as an "applet" adjunct to a document. Once transferred to the user's machine, the JAVA applet can be executed locally, with no need by the reader to acquire any further programs or resources. In a chemical context, the document could be, for example, a set of molecular coordinates, and the JAVA applet could contain instructions to enable three-dimensional rotation of these coordinates within the browser window. It is easy to imagine how a complex chemical diagram could be animated and annotated using a suitably written JAVA applet that would be included with the document. The anticipated reluctance of most users to allow documents of unknown origin to "execute" on their personal computers will need to be addressed, but potentially, JAVA offers a very powerful way of enhancing chemical information.

Automated "image mapping" of one-, two-, and even three-dimensional spectra could become a routine way of helping to associate cross-peaks with particular structural features. This function would in turn require NMR data to be stored in specific formats on WWW servers, together with the definition of appropriate MIME types. The lack of such commonly used format specifications for NMR information would need to be addressed before such a mechanism could be introduced.

Infrared, Rotational, and Mass Spectra

The JCAMP-DX format for the archival of IR spectra has recently been adapted for delivery of native spectral data to the user via hyperlinks.[51] Custom "helper" programs for analysis and display of the data have been developed. As with NMR spectra, mass spectral fragmentation patterns can also be mapped from a two-dimensional diagram into structural coordinates, and the bond cleavage or fragments can be illustrated with the use of markup commands.[46] Similarly, the complex characteristics of a vibrational or rotational spectrum can be greatly clarified by mapping the spectrum to molecular coordinates or displacement coordinates. By this means, the rotational fine structure in a laser fluorescence emission spectrum has been annotated by hyperlinking to an animation of the particular associated normal vibrational mode.[46] This combination of "active" spectral data together with detailed spectral markup holds great promise for the delivery of spectroscopic information.

Chemically Useful Navigation Aids

There has hitherto been little development of aids to navigating chemically specific documents. Some existing tools can, however, be used for this purpose. For example, a complex reaction diagram can be "mapped" to textual explanations of the various individual structures. An image can be defined to contain "hot spots", which may consist of circular or rectangular areas or even irregular polygons. These, and the default area that represents a catch-all for the rest of the diagram, can be defined in an "image map" that contains hyperlinks to URLs that point to either further text, other diagrams, animation, sounds, or network resources such as stored molecular coordinates, databases, and electronic journals.

Figure 12.11 shows how a synthetic reaction scheme[52] can be hyperlinked so that individual intermediates in the sequence or particular functional group transformations can be further elaborated with reagents, explanations behind each step, and perhaps crystallographic coordinates for individual molecules. This mechanism allows a complex synthetic sequence to be "choreographed" to show the logic behind the synthesis and to allow individual steps to be annotated separately. Version 3.0 of HyperText markup language (HTML) is expected to support more elaborate graphical composition that would allow one image to be superimposed over another and permit more subtle integration of text and images on the screen.

In the near future, navigating a complex three-dimensional molecular object will be possible using a new protocol known as VRML or virtual reality modeling language.[48] Whereas HTML is confined to a

Figure 12.11. A synthetic reaction diagram with hyperlinks.

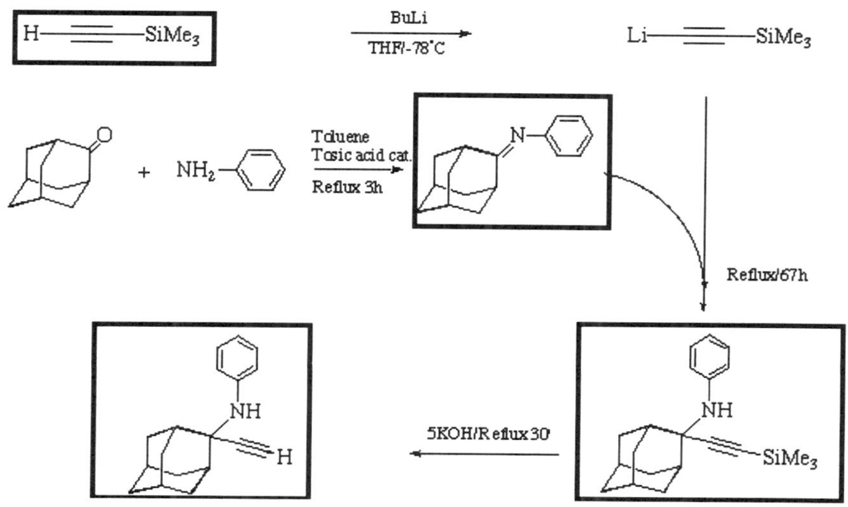

description of two-dimensional objects such as text or images, VRML enables three-dimensional "scenes" to be described. Individual objects in a scene can be associated with hyperlinks via the standard HTTP (HyperText transport protocol) and URL protocol and displayed using a VRML browser such as *WebSpace*. A combination of HTML- and VRML-based browsing on the World Wide Web offers the prospect of a powerful and unique mechanism for molecular navigation.[53]

Complex Reaction Schemes

Traditionally, illustrating a metabolic pathway or a complex synthetic reaction scheme can result in highly symbolic illustrations that require a great deal of annotation and footnotes. The use of image mapping in association with hyperlinks to other documents, diagrams, or molecular coordinates can serve to greatly simplify such information. For example, the Krebs cycle has been so annotated,[46] as has an illustration of the total synthesis of papuamine[52] including linking to appropriate crystallographically derived coordinates, reagents, and descriptions of the logic behind each step in the synthesis. In teaching applications, hyperlinking of reaction schemes for laboratory experiments can be used for the delivery of current and relevant safety and hazard information for each reagent involved.

Chemical Collaboratories

A simple extension to the hyperactive molecule concept would be to allow collaborative visualization of molecular data in which two or more synchronous windows are simultaneously created on remote computer screens associated with collaborators on the project, with a concurrent videoconferencing and audio link.[54] In effect a shared molecular "white board" can be made available, within which molecular data can be manipulated and edited by several people and saved simultaneously at all sites. The WWW in this sense becomes a scientific project-management tool operating on a global scale, particularly as features such as the Common Client Interface[2] become incorporated into chemically cognizant programs.

This area is still in its infancy, being currently limited by the relatively poor networking infrastructure available to most institutions and the almost total lack of chemically oriented software. The gradual introduction of networking technology with fewer so-called latency problems, such as ATM (asynchronous transport mode), and the development of VRML applications suggests significant growth in the future for chemical collaboratories.

The Future for Hyperactive Molecular Data

Many other applications of the concept of hyperactive molecules can be envisaged, such as the provision of crystallographic and instrumental data, tables of numerical information, molecular force fields and numerical basis sets, neural network connection tables, quantitative structure–activity relationship (QSAR) coefficients, symbolic mathematical notation, and other content-rich information. Perhaps the most intriguing association would be with automation and robotics, whereby network-based data could be passed to an instrument or batch synthesis system without necessarily any human intervention.

What are the implications of such a digital environment? One benefit is that the reader could almost instantly evaluate the conclusions presented in a scientific paper or report by having access to the original data. This ability has profound implications for the quality and reproducibility of chemical information. Some concerns have been expressed that distribution mechanisms such as the WWW may result in a proliferation of low quality data, or "noise". We take an opposite point of view here, believing that such instant verification of published data should have the effect of increasing quality!

Nevertheless, significant issues will have to be addressed. First, the sheer volume of the data that could be archived in this way will rapidly lead to a major storage problem. Should the data be stored on the local system or spectrometer where it was originally generated, and where of course it is directly under the control of its originators, or should it be stored in a "trusted" location under the control perhaps of a scientific society or publisher? If the latter, then how are the boundaries of the data defined; that is, to what depth should a hyperlink reference be followed, and should all the referenced documents also be stored? Either way, the cost of such storage may be significant.

If the data is to be stored, then for how long? Given the normal lifetime of most computer systems, it seems unlikely that any commitment to store material for more than a few years could be undertaken. Even long-term archival mechanisms such as CD-ROM are rarely regarded as providing a solution for more than about 20 years. Perhaps the solution is to convert the data to highly structured formats capable of translation into any new standards that emerge over the years.

Another issue relates to the "trusted" nature of the data. It is possible now to encrypt data in such a way that the decryption key is readily available to the scientific community, the purpose being simply to "date-stamp" the information. Such draconian measures may not be necessary for much of deposited data, which might be easily and independently verifiable using suitable algorithms. Undoubtedly however, digital storage and transmission of data could lead to an increase in plagiarism, and disputes over priority in scientific discovery.

One possible application of network-resident data that would depend on this data being readily available and not encrypted is the application of "intelligent agents". If data is accessible via a hyperlink, should we allow it to be searched and indexed by an appropriate program that would roam the Internet seeking particular attributes? Such a molecular agent could be arbitrarily complex; for example one can imagine novel correlations between widely disparate molecular data being discovered. In a sense, the WWW becomes another chemistry research tool, from which original and perhaps even general conclusions can be drawn. One might imagine creating an agent that would roam the Internet analyzing the stereochemical routes of all published examples of pericyclic reactions; a sort of Nobel Prize winning robot!

Viewed as a large and active molecular database from which much commercial advantage could be gained, would it be reasonable to retain the totally free element that characterizes much of the data on the current Internet? Is it acceptable that some types of data might respond with a request to the agent for "e-money" (electronic money)? The agent might need to have a proxy for automatic financial transactions.

Another fine dividing line is to what extent information collated by an agent constitutes new research or scientific plagiarism. Certainly the opportunity to automatically discover inconsistencies in published information and as a consequence to systematically improve the quality of chemical data has to be welcomed.

We are undoubtedly a few years away from such methods of operation, but the first stage of providing readily accessible "live" chemical information in a potentially globally searchable form has clearly been achieved. Problems that still need to be solved are how to index such hyperactive molecular information in a chemically cognizant sense, and how to improve the metaphors for navigating through such information. Currently, a two-dimensional diagram can be mapped and hence navigated in a molecular sense. The next challenge is to map three-dimensional molecular coordinates in a useful sense in order to highlight active sites and other interesting regions. The development of chemical markup languages holds great promise in this area. The reverse process, of associating regions of molecules, or individual atoms and functional groups, with their own uniform resource locators to further information is also currently underway.

Concluding Remarks

In the early 1980s, the use of computer networks in most chemistry departments was left to hardened theoreticians in search of remote machines with which to crunch numbers, and to perhaps the local librarian, information officer, or other specialists. Few users had any sort of network connection to their offices, and the Internet was far from recognized as a *sine qua non* of the laboratory environment. A decade on and it has now acquired a friendly face in most departments, a significant amount of cable, and in some ways the role of a chemical notebook. In no small measure, recent growth has been due to the introduction of the WWW and programs such as `Mosaic` and `Netscape`. Most significantly, we finally have recognition that chemistry responds well not only to text but to graphical and even moving images. Now there are new challenges and issues to confront, including standardization of file formats, a major change in the entire cost structure of scientific information delivery, and an even greater impact on our research and teaching activities.

References

[1]Kopelevich, M., Department of Chemistry, University of California at Los Angeles, **http://www.chem.ucla.edu/chempointers.html**

[2]Thompson, D. "NCSA `Mosaic` Common Client Interface", presented at the WWW Conference at Chicago, IL, October 17–20, 1994. For a specification, see **http://yahoo.ncsa.uiuc.edu/mosaic/cci.spec**

[3]Casher, O.; Rzepa, H. S. *Comput. Graphics* **1995,** *29,* 52.

[4]Chemical Physics preprint service, **http://www.chem.brown.edu/chem-ph.html**

[5]Rzepa, H. S.; Whitaker, B. J.; Winter, M. J. *J. Chem. Soc. Chem. Commun.* **1994,** 1907.

[6]O'Hagan, D.; Rzepa, H. S. *J. Chem. Soc. Chem. Commun.* **1994,** 2058.

[7]*Protein Science;* White, D., Ed.; 1995; **http://www.prosci.uci.edu/**

[8]Richardson, D. C.; Richardson, J. S. *Protein Science* **1992,** *1,* 3; Richardson, D. C.; Richardson, J. S. *Trends Biochem. Sci.* **1994,** *19,* 135.

[9]*J. Mol. Mod.* Clark, T., Ed.; 1995; **http://derioc1.organik.uni-erlangen.de/info/JMOLMOD/jmolinfo.html**

[10]"Electronic Manuscript Preparation and Markup", International Organisation for Standardisation: Geneva, Switzerland; NISO/ANSI/ISO 12083, 1994.

[11]Padwa, A., International Society of Heterocyclic Chemists. See URL **http://euch6f.chem.emory.edu/ishc.html** and **http://www.ch.ic.ac.uk/ishc/**

[12]The WWW94 Geneva Conference proceedings are available via the URL **http://www.elsevier.nl/**

[13]The WWW94 Chicago Conference proceedings are available via the URL **http://www.ncsa.uiuc.edu/SDG/IT94/Agenda/Agenda.html**

[14]Rzepa, H. S. *Comput. Networks ISDN Syst.* **1994,** *27,* 317.

[15]Rzepa, H. S. *ISDN Comput. Network Syst.* **1994,** 384; **http://www.ch.ic.ac.uk/talks/www94_chem_ws_report.html**

[16]Krieger, J. H.; Illman, D. L. *Chem. Eng. News* December 12, 1994, p 29; Bachrach, S. M. *J. Comput. Info Comput. Sci.* **1995,** *35;* 431; URL **http://hackberry.chem.niu.edu/ECCC2/ECCCIntro.html**

[17]*Proceedings of the First Electronic Computational Chemistry Conference, Nov. 7–Dec. 2, 1994;* Bachrach, S. M.; Boyd, D. B.; Gray, S. K.; Hase, W.; Rzepa, H. S., Eds.; ARInternet: Landover, MD, 1995.

[18]Hughes, K., **http://www.eit.com/software/hypermail/hypermail.html**

[19]Bachrach, S. M.; Pierce, T. H.; Rzepa, H. S.; 210th ACS National Meeting, Chicago, IL, August 1995; URL **http://hackberry.chem.niu.edu/Infobahn/Paper38**

[20] "First Electronic Conference on Trends in Organic Chemistry"; Rzepa, H. S.; Goodman, J. M., Eds.; June 12–23, 1995; **http://www.ch.ic.ac.uk/ectoc/**

[21]Rzepa, H. S., **http://www.ch.ic.ac.uk/talks/**

[22]Further information is given at URL **http://www.ceneca.com/**

[23]Peitsch, M. C., **http://expasy.hcuge.ch/pub/Graphics/IMAGES/GIF**

[24]Laaksonen, L., Center for Supercomputing Applications, Espoo, Finland, **http://www.csc.fi/lul/chem/graphics.html**

[25]Savary, F.; Weber, J., Department of Chemistry, University of Geneva; Geneva, Switzerland, **http://scsg9.unige.ch/eng/toc.html**

[26]Chnel, J. S-F., URL **http://www.imb-jena.de/IMAGE.html**

[27]Wong, Y.-L.; Yip, C.-W., URL **http://www.chem.vt.edu/yip/VR/3DAO.html**

[28]Bachrach, S. M., Department of Chemistry, Northern Illinois University, **http://hackberry.chem.niu.edu:70/0/webpage.html**

[29]Fitzgerald, P. C., National Institutes of Health, **http://www.nih.gov/htbin/pdb**

[30]Murzin, A.; Brenner, S. E.; Hubbard, T. J. P.; Chotia, C. "The SCOP Project: Structural Classification of Proteins"; MRC Laboratory of Molecular Biology and Centre for Protein Engineering: Cambridge, England, **http://www.bio.cam.ac.uk:80/scop/**

[31]Schuchardt, K.; Feller, D., **http://www.emsl.pnl.gov:2080/forms/basis-form.html**

[32]Sadowski, J.; Gasteiger, J. *Chem. Rev.* **1993,** *93,* 2567.

[33]Sadowski, J.; Gasteiger, J.; Klebe, G. *J. Chem. Inf. Comput. Sci.* **1994,** *34,* 1000.

[34]See also **http://schiele.organik.uni-erlangen.de/services/3d.html**

[35]Dunford-Shore, B. H.; Fabrizio, F.; Holcomb, J.; Wise, W.; Kazic, T. "Klotho Project: Biochemical Compounds Declarative Database", Institute for Biomedical Computing, Washington University: St. Louis, MO, **http://ibc.wustl.edu/klotho/**

[36]URL **http://www.daylight.com/kisstxt/daylight-booth.html**

[37]URL **http://www.silverplatter.com/sampler/webspirs.html**

[38]Rzepa, H. S., URL **http://www.ch.ic.ac.uk/e-chemlib/**

[39]Winter, M. J., URL **http://www.shef.ac.uk:80/~chem/web-elements/**
[40]Winter, M. J.; Rzepa, H. S.; Whitaker, B. J. *Chem. Br.* **1995,** *31,* 685.
[41]Rzepa, H. S., **http://www.ch.ic.ac.uk/igcoc/**
[42]The GlobeWide Network Academy, **http://uu-nna.mit.edu:8001/uu-gna/index.html**
[43]For the Biological MOO see **http://bioinformatics.weizmann.ac.il:70/1s/biomoo**
[44]Murray-Rust, P.; Mills, A. "VSNS-PPS Course on the Principles of Protein Structure", Birkbeck College, London, 1994; **http://www.cryst.bbk.ac.uk/PPS/index.html**
[45]Leach, C. J.; Rzepa, H. S. In *Proceedings of the First Electronic Computational Chemistry Conference, Nov. 7–Dec. 2, 1994;* Bachrach, S. M.; Boyd, D. B.; Gray, S. K.; Hase, W.; Rzepa, H. S., Eds.; ARInternet: Landover, MD, 1995.
[46]Casher, O.; Chandramohan, G.; Hargreaves, M.; Murray-Rust, P.; Sayle, R.; Rzepa, H. S.; Whitaker, B. J. *J. Chem. Soc. Perkin Trans. 2* **1995,** 7.
[47] Sayle, R. "`RasMol`: A Molecular Visualisation System. V2.5 (1994)", Glaxo Research and Development, Greenford, United Kingdom.
[48]Pesce, M.; Bell, G.; Parisi, A. "VRML 1.0". See also **http://www.wired.com/vrml, http://www.eit.com/vrml** and **http://www.sd.tgs.com/~template/** For a collection of molecular entities encoded in VRML, see Rzepa, H. S.; Casher, O., **http://www.ch.ic.ac.uk/VRML/**
[49]Murray-Rust, P.; Leach, C. J.; Rzepa, H. S., URL **http://www.ch.ic.ac.uk/cml/**
[50]URLs **http://java.sun.com/** and **http://java.sun.com/applets/applets/ChemicalModels/index.html**
[51]Lancashire, R. J.; Department of Chemistry, University of West Indies: Mona, Kingston, Jamaica, **http://wwwchem.uwimona.edu.jm:1104/**
[52]Barrett, A. G. M.; Boys, M. L.; Boehm, T. *J. Chem. Soc. Chem. Commun.* **1994,** 1881.
[53]Casher, O.; Rzepa, H. S. *ACS Meeting Abstracts, Anaheim, CA, April 1995;* American Chemical Society: Washington, DC. See **http://www.ch.ic.ac.uk/ACS/anaheim/**
[54]Cacher, O.; Pilling, M.; Rzepa, H. S.; Whitaker, B. J., SuperJanet ATM project, demonstrated on July 13, 1994, between Imperial College, London, and Leeds University.

chapter thirteen

Chemical Industry and the Internet

T. H. Pierce and T. J. Cozzolino

In This Chapter

This chapter addresses the use of the Internet in the chemical industry and the shift from its use as a special network to its use as a commodity network. Several practical uses for the Internet and Internet-related technologies are presented. Some of the major challenges for industry resulting from the use of the commercial Internet are reviewed, including costs and security, education and training, and efficiency and effectiveness. Finally, the paradigm shift toward distributed information and the future of world-wide networks and the chemical industry are discussed.

3223–7/96/0277$37.00/1

BEFORE THE EARLY 1990s, AN INTERNET CONNECTION was something of a rarity, a luxury that only U.S. governmental institutions, universities, or defense contractors valued. However, in the past few years, commercial use of the network has increased dramatically. So much so, in fact, that commercial users of the Internet now outnumber those in academia. How did this happen? More important, what is so useful and significant about connecting to this "network of networks"? In this chapter we will examine the recent growth rates of the Internet, the major challenges posed to companies considering a connection, and the special contributions made to the growth of the Internet by the chemical industry. Finally, we will consider future uses of global networks and their impact on scientific and commercial collaboration.

Special Forces and Convergence

"The Internet was born about 25 years ago, as a U.S. Defense Department network called the ARPANET [Advanced Research Projects Agency Network]. The ARPANET was an experimental network designed to support military research—in particular, research about how to build networks that could withstand partial outages (like bomb attacks) and still function".[1] By the 1990s, however, several forces that began to converge rapidly escalated the use and importance of the Internet. The first was the growth of the Internet regional provider. More than 50 commercial companies sell Internet connectivity. Prices vary from as little as $20 per month for simple dial-up connectivity to more than $5000 per month for high-speed (T-1) service.

Internet connectivity has become a commodity, a common service available to anyone who wishes to be connected. Common "gateways" between providers, such as America Online, CompuServe, and Prodigy, and the Internet have become commonplace, and this situation further allows "at home" customers to become exposed to global networks. Currently, this market is expanding rapidly, as providers seek to offer cost-effective connectivity. Regional providers and some telecommunications companies are also beginning to offer ISDN (Integrated Services Digital Network) service to the home consumer. ISDN promises near-industrial-speed connectivity for as little as $30 per month. Clearly, Internet connectivity is within the grasp of every commercial company and most home computer users.

Another important development was the formation of the Commercial Internet Exchange (CIX) in March 1991. As more and more commercial enterprises began using the Internet, a severe problem surfaced: The original National Science Foundation Network (NSFNET) restricted communications to noncommercial traffic. This restriction prohibited the transmittal of information between commercial companies and thus severely decreased the value of their connection. Regional providers join the CIX, a nonprofit trade organization, which enables them to handle "all types of traffic (as long as it does not violate applicable laws)—without requiring their users to route commercial traffic differently from noncommercial traffic (research and education). When using CIX member networks, all company traffic goes over the same network without fear of violating NSFNET or Internet acceptable-use policies. Network managers need maintain only one external network, which connects to a CIX member."[2] For the first time, commercial communications and even simple financial transactions were allowed over the Internet. Such a development was crucial to the sustained growth of the Internet for these reasons:

- It continued to add more commercial users to the network.
- It reduced end-user costs by fostering competition.
- It fueled interest by corporations in developing network tools.
- It allowed more people to make money by using the network.

As the U.S. government reduced its funding of the original NSFNET, it also began a major plan to build a National Research and Education Network (NREN). This initiative grew into the National Information Infrastructure (NII), calling for legislation to "increase competition in communications markets by explicitly promoting private sector infrastructure investments—both by companies already in the market and those seeking entry."[3] It is now a "good thing" to use the global Internet for commerce. For those used to the acceptable-use policies of the 1970s and 1980s, this idea is quite a policy shift.

The development of new hardware and software technologies has also contributed significantly to the ability of the Internet to deliver on its promise of rapid information delivery. Communications devices such as the router, intelligent hub, and the once-lowly modem deliver performance at levels several orders of magnitude greater than what was possible only 2 years ago. Emerging communications protocols such as asynchronous transfer mode (ATM) and

ISDN promise ever faster information transfer between desktops at home and the office.

Software is also evolving at a rapid pace. Most notable are the Internet "discovery tools", which are Gopher, Wide Area Information Service (WAIS), and the World Wide Web (WWW or the Web). These tools present the chaotic Internet via metaphors that most computer users are comfortable with. Gopher shows portions of the network as a set of menus, WAIS as a group of databases, and WWW as a series of HyperText "pages" (Figure 13.1). This presentation frees the user to explore (or "surf") the network to find new resources while actually getting productive work done. The growth in the use of these tools has been staggering, as shown in Figure 13.2 (WWW usage, for example, is growing at over 1000% annually).

A final force not to be underestimated is that of the mass media. Before 1985, one would be hard-pressed to find any reference to computer networks in the major media (newspapers, news magazines, and television news). However, as seen in Figure 13.3, this situation is no longer true. Major publications such as *Time*, *Newsweek*, and *The New York Times* have all featured prominent articles describing aspects of the

Figure 13.1. Different presentations of data by various Internet discovery tools.

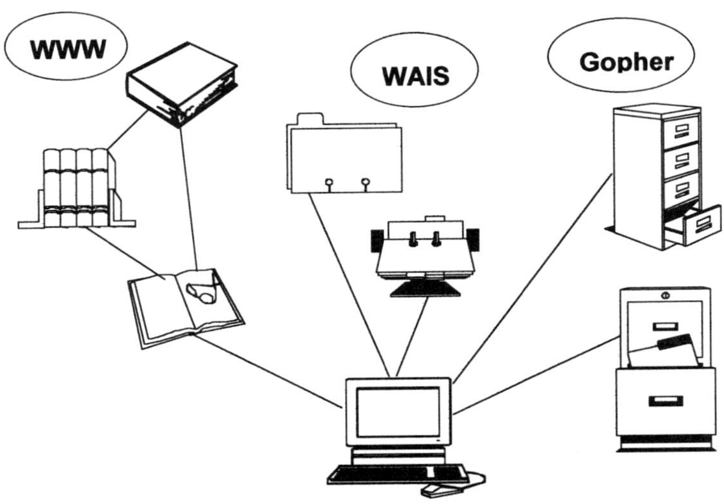

Figure 13.2. WWW and Gopher growth. (Source nic.merit.edu).

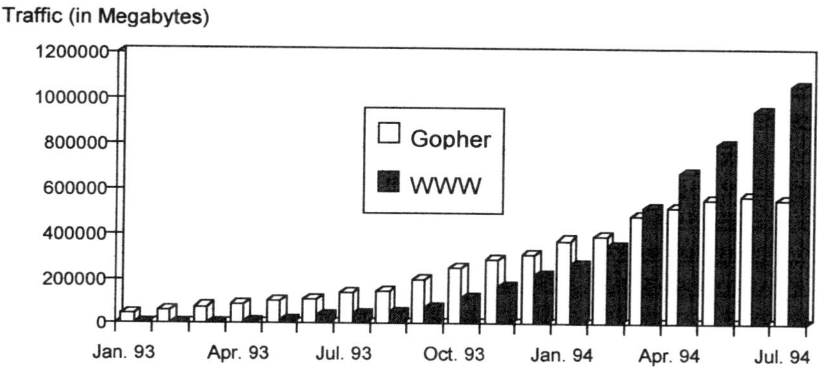

Figure 13.3. Internet tools in the printed media (article counts as of September 1, 1994.)

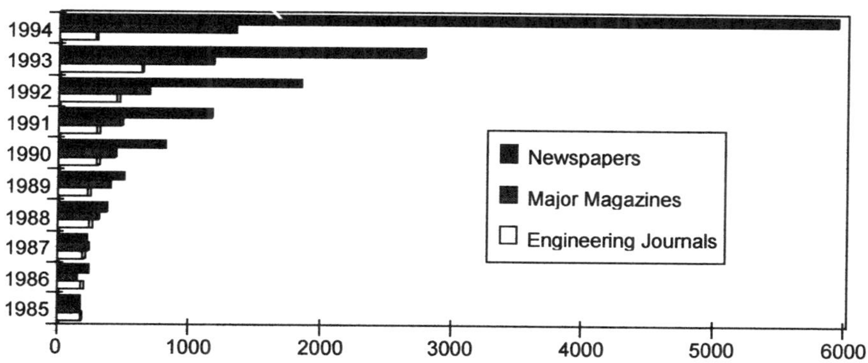

Internet and Internet-related technologies. The Internet truly has become part of our culture.

We believe that this convergence of availability, permission, technology, and governmental and societal forces has given rise to an Internet that is drastically different from the original experiment conducted by the Department of Defense in the 1970s. In fact, this new commercial Internet could not have been born without each of these forces. Without the CIX and the effective obliteration of the acceptable-use policy (AUP), traffic would still be limited to noncommercial uses, and this limitation would discourage innovation from for-profit businesses. If the telecommunications infrastructure were not improving at its current rate, backbones would be saturated, and this condition would frustrate new or potential users and dampen interest. And without the NII and societal forces at work, the Internet would still be a curiosity, instead of an opportunity to create new information-based industries as well as improve technical and business communications. Truly, now is the time to explore and take advantage of the Internet.

A Commercial Journey down the Information Highway

The chemical industry has a long history of advanced computer use. In the 1970s, the use of molecular modeling became prevalent in the pharmaceutical industries and spread to petrochemical and specialty chemical makers. As molecular modeling use grew, so did the complexity of the systems studied. Typical molecular modeling groups quickly outgrew in-house computing facilities and began to look to the national supercomputer centers to meet their needs for high-quality quantum mechanics calculations. This need led to close collaboration of the chemical industry with the national supercomputer centers and an increased use of the NSFNET. Although the connection was originally used for direct access to the supercomputer centers, researchers quickly learned the utility of world-wide electronic mail (e-mail), file-transfer protocol (FTP) sites, and other collaborative information sharing.

This exact path was traveled by the Rohm and Haas Company. In the late 1970s the use of molecular modeling in the Agricultural Chemicals Research group was common.[4] Soon the scope of the problems to be undertaken exceeded the in-house network of workstations and computer servers. This situation led to a 1989 research effort to connect to the Minnesota Supercomputer Center to perform high-quality quantum mechanics calculations and fluid dynamics simulations. We soon learned that this connection could also be used to access the Internet for other purposes. After several months, a modest amount of internal publicity described the uses of the new Internet connection.

The Internet was beginning to become popular in the mass media, and so other individuals in the corporation began to take advantage of it for e-mail and other simple uses. Internally, we began using the

Usenet bulletin board system via the NETNEWS application on our IBM VM mainframe and offered an e-mail gateway between our corporate IBM PROFS (professional office system) and the Internet's simple mail-transfer protocol (SMTP) mail.

The combination of NETNEWS and e-mail proved to be synergistic. Individuals discovered the Internet through two paths. Many needed to send e-mail to colleagues whom they met at conferences or found an e-mail address in a trade magazine. These individuals then inquired about using e-mail and soon became users.

Others heard about or discovered NETNEWS. This version of Usenet allowed individuals to browse the newsgroups and discover other resources on the Internet. NETNEWS was still a prototype without official support by our organization, so we included a disclaimer on the banner page to inform users of its nature and the fact that our company had no control over its contents. Soon, we began to create our own internal-use newsgroups, which we call **rh** newsgroups, to discuss internal information technology (I/T), project management, and internal technical issues.

Usage grew by word of mouth for 2 years. Originally we had 10 users who accessed the remote supercomputers in Minnesota. As desktop UNIX workstations grew more powerful, our use of external supercomputers declined. At the same time, our use of the Internet grew dramatically. By 1992, hundreds of internal users were taking advantage of our Internet connection and using it in many business areas.

Another development was that requests for internal support were coming from people throughout our corporation. To investigate the uses of the Internet and to determine what internal support requirements might be needed, we formed a project team. The charter of this committee was to achieve the following goals:

- discover current Internet usage by Rohm and Haas and other chemical companies
- orient our internal management on Internet resources
- identify potential Internet usage by the Research Division
- review potential support issues with internal management
- suggest ways to address the support issues

We gathered data from four major sources: an external survey of Internet use by other companies, an external literature search, internal statistics regarding use of Internet applications, and an internal survey of selected Rohm and Haas employees.

The External Survey

Our external survey's purpose was to gather information about current and potential Internet usage in the chemical industry. To obtain this information, we posted a note to several newsgroups via Usenet and selected mailing lists asking the following questions:

- Is anyone willing to share their experiences, successes, and usage patterns regarding the Internet?
- What are the value-added benefits of using the Internet for your research and business?
- What are the biggest challenges facing you, the providers of Internet services?
- How difficult has it been to change your particular corporate or information technology information services cultures to embrace this "brave new world" view?"

This posting was sent to the following newsgroups and mailing lists:

Name	*Type of Information Service*	*Audience*
alt.internet.services	Usenet	Internet users
bit.listserv.chminf-l	Bitnet mailing list	chemical industry librarians
bit.listserv.libref-l	Bitnet mailing list	reference librarians
bit.listserv.pacs-l	Bitnet mailing list	information systems users
comp.infosystems.gopher	Usenet	Gopher users
comp.infosystems.wais	Usenet	WAIS users
sci.chem	Usenet	chemists
sci.engr.chem	Usenet	chemical engineers
chemistry@osc.edu	mailing list	computational chemists

We received 65 responses to this survey. Thirty-two were from commercial corporations. We found that the responses from both educational and commercial sites were very similar. The Appendix presents a brief statistical breakdown of the survey data, as well as a list of the respondents.

By far, the largest use of the Internet is in the area of e-mail. Both commercial and academic institutions depend on Internet-based mail to communicate with colleagues, customers, and suppliers. Historically, e-mail has been one of the driving forces of Internet use. One respondent said that it is "the most universal useful service. ... [e-mail makes it] easier to pull together working groups on particular topics or to collect and disseminate information."

Surprisingly, of perhaps greater importance is the use of bulletin boards (Usenet) and mailing lists. Often, Internet customers start with a news feed, by monitoring newsgroups of interest and beginning to understand what the Internet is all about. These groups educate readers about general issues through frequently asked questions (FAQs) lists and by discussing the experiences and solutions to various issues that are raised. Usenet newsgroups discuss folklore, share experiences, and provide pointers to software needed to solve specific problems. Generally, the software is available from anonymous FTP sites, and anyone with an Internet connection can access it. As users progress to posting questions and answers to the newsgroups, they begin to meet the people behind the keyboards, and they progress to closer collaboration and personal e-mail.

One respondent said that "Usenet news has resulted in several useful techniques and ideas coming to our attention. ... With a good newsreader and some filtering, it is a good way to pick up information and help." Another respondent begins his research by querying other researchers via Usenet for references, suggestions, or software that may be applicable. "To date I have never been disappointed. ... Usually one or two of the responses are exactly what I need to make progress on my research."

Nearly a third of the respondents stated that they carry out some kind of collaborative research or work with colleagues over the Internet. Symposia and seminars are held, and other collective work is carried out, as is the easy movement of documents, results, etc. In fact, some projects could not have happened without the Internet because it allowed fast and inexpensive communication with researchers in other parts of the world. "The ability to communicate directly with technical professionals in other enterprises and to have access to the user forums on the Internet" is invaluable, as one respondent stated.

A large number of those polled use the Internet as their basis of connection to commercial services and databases, such as Dialog and Chemical Abstracts Service. Users find that an Internet-based connection is faster, more robust, and easier to use than modems. Most likely, in the future more companies will depend on up-to-the-minute information (continuous education) that will be delivered via Internet and Internet-like services. One respondent appreciated the ability to access the "vast amount of reference data including on-line catalogs of major universities." Finally, most of those people who have used the Internet quickly realize it can be a vital part of getting their jobs done. Another respondent appreciated the ability to send emergency patches over the network by e-mail or by FTP and the ability to receive electronic copies of files that are giving users problems and try them out on his own computer. This approach makes computer support much easier, faster, and more efficient. A graphical summary of the uses by researchers outside of Rohm and Haas appears in Figure 13.4.

Other information gathered from the external survey dealt with the concerns that corporations and universities have regarding their Internet connections (*see* Figure 13.5). The three main concerns were security, internal management and policy issues, and education and support of new Internet users. The Internet is seen as a potential security risk for large corporations. Often, the external Internet connection began as a research project, and then became popular enough to warrant a migration to "production" status. This step comes as a "culture shock" to traditional computer-support organizations (management information systems or MIS departments) who are more comfortable with secure environments that are under their control.

Another major issue is the "time sink" nature of Internet usage. With the vast amount of on-line services available, new users may feel over-

Figure 13.4. Summary of the Internet uses by researchers outside of Rohm and Haas.

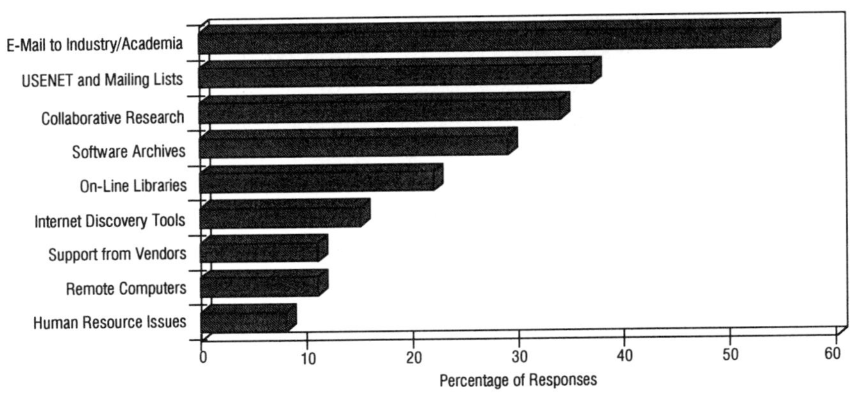

Figure 13.5. Summary of the concerns of Internet users outside of Rohm and Haas.

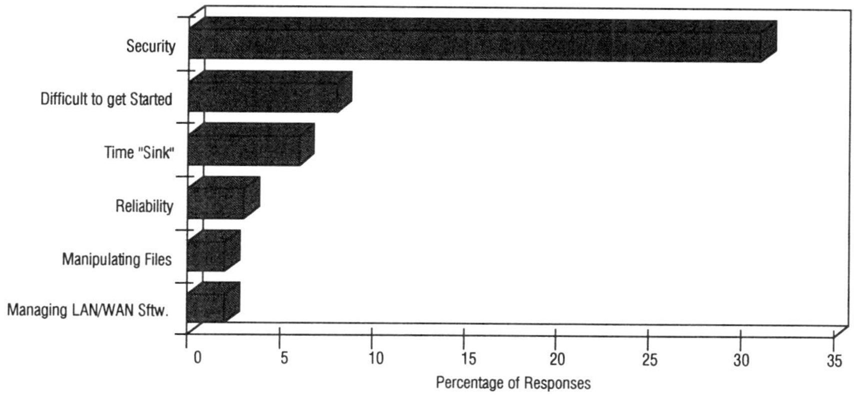

whelmed and uncertain of where to start. Most people may not realize that learning how to effectively navigate a world-wide network simply takes time. Often, users can get "lost in cyberspace". The Internet also changes so quickly that people feel compelled to "relearn" how to use new resources and new tools. These issues will subside with time.

The "collision" between the Internet (open, seen as unregulated) and internal MIS (under tighter control) is a significant obstacle to the acceptance of the Internet in large organizations. In fact, this culture clash has in some cases led to delays in Internet deployment. Even when the Internet is accepted within large institutions, however, the task is far from over. Much support and training work is still to be done. Internet discovery tools (Archie, Gopher, WAIS, and WWW) seem to be a good first step in solving the "information overload" problem. With more than 10 Terabytes of information to choose from, we believe that these kinds of tools are essential as the Internet continues to grow at its current explosive rate.

The Internal Survey

To determine why Rohm and Haas researchers were using the Internet, we measured the traffic passing through our Internet gateway and the frequency of Usenet usage by individuals. We also counted the number of TCP/IP (transmission control protocol/Internet protocol) hosts in the three-site Delaware Valley area. Our work was completed over a 6-month period during 1993.

Our internal survey sample, though small, gave us a good snapshot of Internet users, from novices to experts. Each member of our project team selected a novice, an intermediate, and an expert user. This selection process gave us a good cross-section of internal customers with whom to conduct our interviews. The interviews lasted 1–2 hours and were attended by at most four interviewees. The respondents represented a number of functions and businesses within the Rohm and Haas Company. The results of our internal study are shown in Figure 13.6.

Our hope was to gauge how our internal customers were using the network and to help us plan for future training, development, and support infrastructure. Specifically, we hoped to learn:

- the types of services used
- how access is used
- what fraction of time spent is added value
- business uses of the Internet
- the types of training needed
- missing items: what products and services are not provided?

We asked the following questions:

- *What do you use the Internet for, and how does it help your job?* All those interviewed used the Internet for e-mail and had varying degrees of

Figure 13.6. Summary of the Internet uses by Rohm and Haas employees.

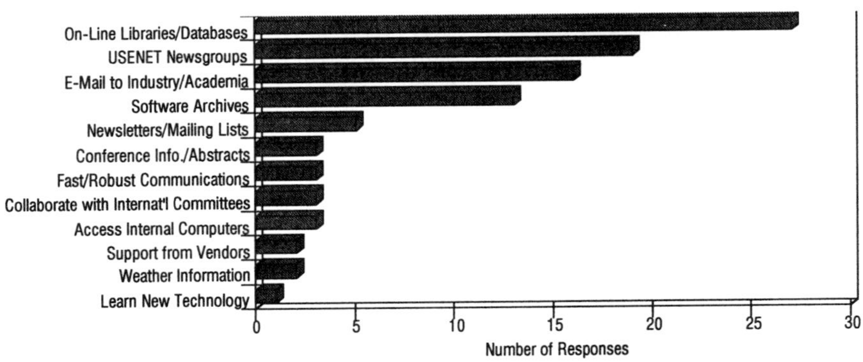

difficulty with it. Many were confused as to where the problems occurred. For example, if Internet mail was "bounced", or was not delivered, users were unsure if the problem originated on the local mail system, on the remote host, or somewhere in-between.

■ *How did you learn to use the Internet?* Most people learned about the Internet from internal informal discussions (55%); however a large percentage (45%) learned about the Internet from outside sources, including external conferences and previous educational or job experience. People were also beginning to share Internet information among themselves. This fact partially explains the large increase in the Internet information requests from Research, Corporate, and other Rohm and Haas locations.

■ *What kinds of problems have you encountered in using the Internet?* The most-often described problem with using the Internet was the lack of documentation on how to use it. Many people talked about how difficult it was to find the information they wanted. Even if they had heard from others that this information was available, they did not know how to find it. They wanted a support person or group to be available to help find information and to maintain a repository of useful Internet databases and software sites. People expressed interest in just-in-time training on applications such as Telnet, FTP, e-mail, and Gopher that was specific to the computer they mainly used. Most people said that if something was not working quite right, they did not know if they had done something wrong or if the computer network was not functional, or perhaps the remote computer was simply unavailable.

■ *What would you like to do with Internet access?* Our internal customers requested a great number of interesting and useful Internet services, including easy-to-remember e-mail addresses for their business cards; more training regarding external database access; electronic data interchange (EDI) using the Internet; electronic access to books-in-print; PC- and Mac-based programs to Usenet, FTP, and e-mail; access to for-pay services like Dow Jones and Lexis/Nexis; and creation of **rh** internal newsgroups for electronic conferencing. One interesting idea suggested by two of the interview groups was the ability to search the Rohm and Haas computer network in much the same way they explore the Internet. For example, Rohm and Haas personnel could perform a Gopher/WAIS search to find useful internal information regarding information services, support procedures, etc.

■ *Is there anything on the Internet that others in Rohm and Haas should know about or use?* The most common statement was that the Internet is for everyone. The caveat that most agreed to was that currently the Internet has the most to offer to people who support and use computers. However, many pointed out that more data for business customers is becoming available now, and this is the fastest growing area of the Internet. Interviewees also believed that there is high potential for Internet usage outside of research, especially in the areas of customer contacts and marketing.

Major Lessons Learned from the Journey

Most of our experiences in using the Internet can be grouped into three major areas: communication, collaboration, and continuous education.

Initially the Internet was used solely to send and receive electronic mail. At that point, users knew only the e-mail address of the recipient. This communication continued to grow and deepen as the customers increased their Internet usage.

The next stage was to read the Usenet bulletin boards. Users perused topics of interest and discovered the availability of a database or software pertaining to their interests. Users then connected to remote sites via FTP or Telnet to retrieve files or access databases. This was the beginning of collaboration.

The third stage of Internet use concerns finding other information on the Internet. At this stage, users learned about Internet discovery tools (Gopher, World Wide Web, WAIS, etc.) to deal with the huge amount of disparate information available. The capability for continuous education and quick retrieval of information can lead to significant technical and business benefits.

We were also struck by the similarities of the Rohm and Haas internal network and the world-wide Internet network. The internal network was a miniature Internet, complete with heterogeneous computers; differing skill levels of users; and diverse uses and interests in chemistry, computer, and business topics. Consequently, the technologies that succeed on the Internet may be transferable to our internal network. The Internet "tests" technologies for file transfer (FTP), information publication and dissemination (Gopher, WAIS, and WWW), and document storage and retrieval (e-mail, FTP, Gopher, WAIS, and WWW). These tools have "survived" and prospered, so using these tools internally as well allows us to make good choices and gives us a high possibly of success when providing similar services to our internal customers. Truly, scalability on the Internet implies scalability to our company.

Practical Applications for Today

Rohm and Haas has been connected to the Internet for the past few years, and has received many benefits from this connection. But there was always a high degree of complexity to anyone attempting to use the Internet to help them in their work. For example, using e-mail required some knowledge about addressing schemes (for example, T. H. Pierce's e-mail address is **THPierce@RohmHaas.Com**) and how to handle "bounced" mail. To connect to remote databases, users had to become confident with the Telnet program, varying login sequences, and multiple terminal types. To retrieve files, users had to find a program of interest and then learn the rules of FTP and binary file transfer. And even then, prerequisite knowledge about file compression was necessary. To our internal I/T service providers, it seemed that the Internet, though incredibly useful, was just too complex to benefit many of our busy colleagues. However, we then discovered and began to use `Mosaic` and other powerful WWW browsers.

To our internal customers, desktop programs like `Mosaic` greatly reduce the need to know the arcane rules of Telnet, FTP, and most other Internet services. Although still far from "shrink wrapped" (that is, purchased in a store), `Mosaic` and its relatives combine common Internet resources in a straightforward and easy-to-understand way. Figure 13.7 shows a chemistry-related WWW page that we have provided for our research users. Newsgroups, Gopher sites, FTP sites, and other WWW sites are all presented in a clear and concise way. The user simply single-clicks on items of interest, and `Mosaic` retrieves them. There is still an occasional trick to learn, but most are related to initial configuration (graphics viewers, TCP/IP communications stack, etc.). Once users have seen `Mosaic` in action, they almost unanimously request it for their own desktop computers, as well as for their computers at home.

Practical business benefits may be derived from the three major uses of the Internet discussed earlier. Business communications involves direct contact between customers and suppliers. The Internet then becomes a basic marketing tool where interested purchasers may

Figure 13.7. Chemistry home page on an internal Web site.

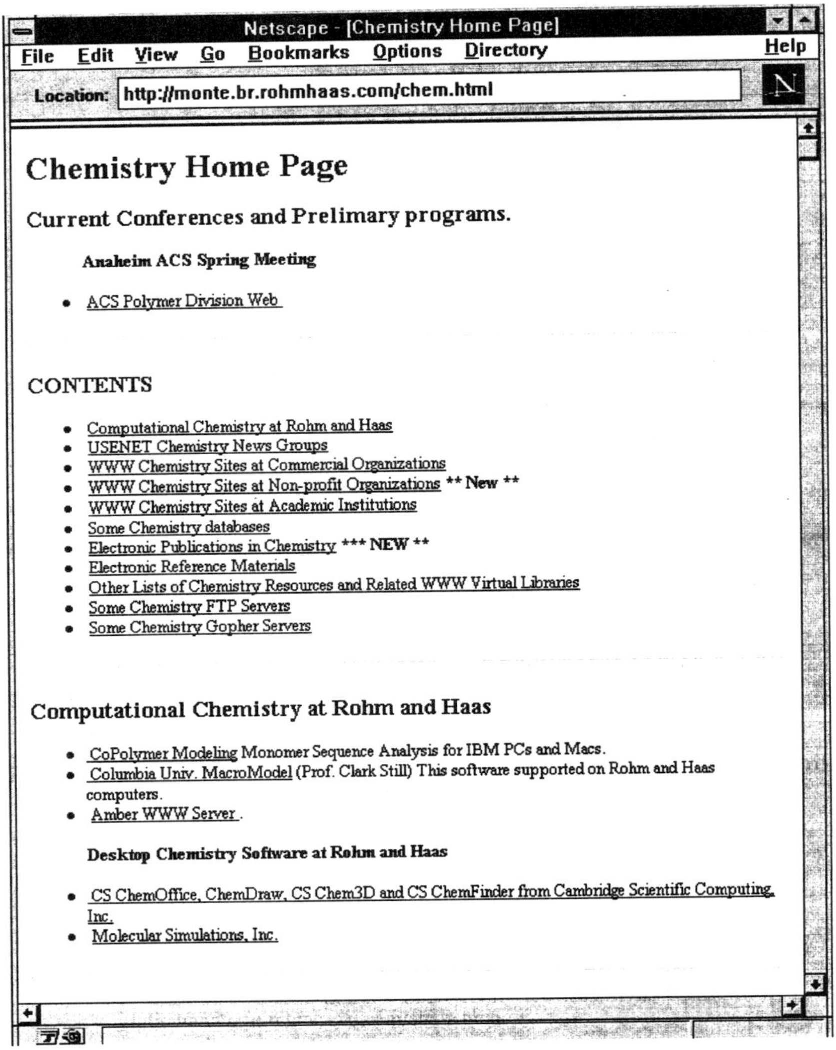

Figure 13.7. Continued. More of the chemistry home page.

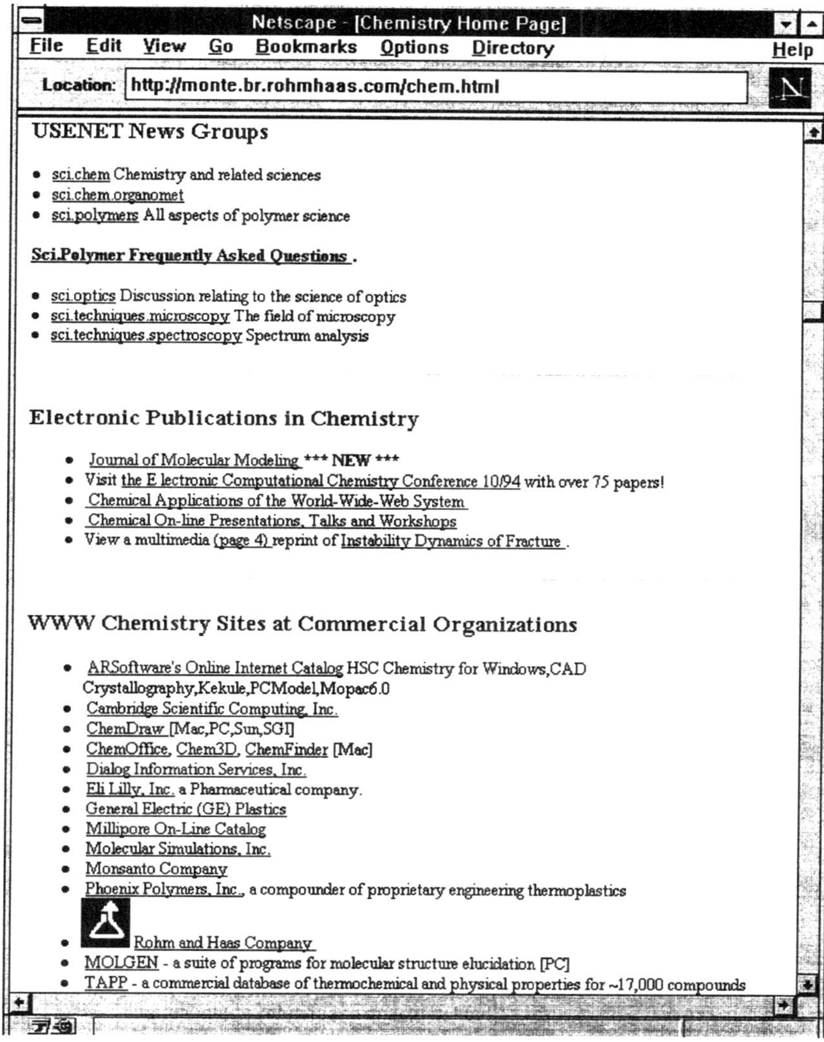

Netscape - [Chemistry Home Page]

File Edit View Go Bookmarks Options Directory Help

Location: http://monte.br.rohmhaas.com/chem.html

USENET News Groups

- sci.chem Chemistry and related sciences
- sci.chem.organomet
- sci.polymers All aspects of polymer science

Sci.Polymer Frequently Asked Questions .

- sci.optics Discussion relating to the science of optics
- sci.techniques.microscopy The field of microscopy
- sci.techniques.spectroscopy Spectrum analysis

Electronic Publications in Chemistry

- Journal of Molecular Modeling *** **NEW** ***
- Visit the E lectronic Computational Chemistry Conference 10/94 with over 75 papers!
- Chemical Applications of the World-Wide-Web System
- Chemical On-line Presentations, Talks and Workshops
- View a multimedia (page 4) reprint of Instability Dynamics of Fracture .

WWW Chemistry Sites at Commercial Organizations

- ARSoftware's Online Internet Catalog HSC Chemistry for Windows,CAD Crystallography,Kekule,PCModel,Mopac6.0
- Cambridge Scientific Computing, Inc.
- ChemDraw [Mac,PC,Sun,SGI]
- ChemOffice, Chem3D, ChemFinder [Mac]
- Dialog Information Services, Inc.
- Eli Lilly, Inc. a Pharmaceutical company.
- General Electric (GE) Plastics
- Millipore On-Line Catalog
- Molecular Simulations, Inc.
- Monsanto Company
- Phoenix Polymers, Inc., a compounder of proprietary engineering thermoplastics
- Rohm and Haas Company
- MOLGEN - a suite of programs for molecular structure elucidation [PC]
- TAPP - a commercial database of thermochemical and physical properties for ~17,000 compounds

find information and easily contact suppliers to answer more specific questions. This accessibility leads to more satisfied and informed customers. Examples are seen in WWW servers produced by commercial companies such as General Electric, Millipore, and Rohm and Haas.

Collaboration in business is seen in partnering efforts between companies in joint ventures, joint marketing efforts, and co-development of products with academic research centers. The Internet's global reach allows timely resolution of problems and issues with a minimum of delay, while allowing the collaborators to share text, graphics, video, and sound. This spectrum of information greatly enhances teamwork over more limited methods using telephones and fax machines.

Businesses may use the Internet to both keep their internal personnel educated in new technologies and fields of interest and to educate the customers in appropriate products and their applications. Usenet is a good demonstration of an Internet application that can educate both internal and external customers. Many computer-related companies, for example, use the Internet technologies such as Usenet, mailing lists, and WWW servers to share support information, bug fixes, and new release information directly with interested customers.

Mine Field or New Frontier?

We have explored some of the history of the Internet and have discussed some practical present-day applications, but considerable challenges remain with managing and using an Internet connection. To many, the Internet is a "mine field" populated with dangerous hackers ready to attack vulnerable machines and corporate information. To others, it is a "new frontier" of opportunity and nearly unbridled freedom. The truth lies somewhere between these two extremes. We will explore some of the major challenges encountered when a company acquires an Internet connection. We encourage you to seek out more complete sources of information about these challenges, and decide for yourselves which cliché is more to your liking.

Security Issues

Any institution that is considering an Internet connection should weigh the risks and benefits and be prudent in its policy:

> A *security policy* is a set of decisions that, collectively, determines an organization's posture toward security. More precisely, a security policy determines the limits of acceptable behavior, and what the response to violations should be. Naturally, security policies will differ from organization to organization. But every organization should have one, if only to let it take action when unacceptable events occur.[5]

The important point here is a that a policy should exist for an organization. Historically, many large companies become connected to the Internet as part of a research project. As the number of Internet users grows to other areas of the corporation, however, the need for a policy becomes quite apparent. At the Intel Corporation, for example, this transition was a painful one: "Our policy was incredibly naive. We did not think it through in depth and did not realize how easy it would be for intruders to exploit gaping holes. Furthermore, we did not have buy-in to our policy. System administrators weren't comfortable with it. Even worse, they were uncomfortable with a policy they couldn't even read. Things had to change."[6]

In considering a policy, a basic question surfaces: What will the strategy be at the connection point, that is, where the Internet and the corporation's physical network connections meet? This strategy can be handled in a number of ways. One is a simple router-based firewall (Figure 13.8A) that filters packets between the internal and external networks. Although simple to implement, this strategy depends on the correct definition of filters and the integrity of the router's software. One major weakness is that little or no logging is inherent in such a firewall router, which means that a corporation would have no way to monitor the number of attempts at penetration and would know of a security breach only after one had taken place.

Another alternative is a series of routers and gateway machines (Figure 13.8B). This arrangement directs all communications through the gateway machine, usually a UNIX or other multiuser computer. The gateway hosts may be configured to log all transactions, providing extensive information about both legitimate usage and illegitimate intrusion attempts. Several commercial implementations of host-based firewalls are available, as well as sources of information about firewalls: such as Cheswick and Bellovin, CERT, at **ftp: ftp.cert.org**, the **firewalls** mailing list; and the **comp.security.***, **alt.security**, and **sci.crypt** newsgroups. Whatever firewall strategy is chosen, it should

Figure 13.8. A: Router-based firewall. B: Series of routers and gateway machines.

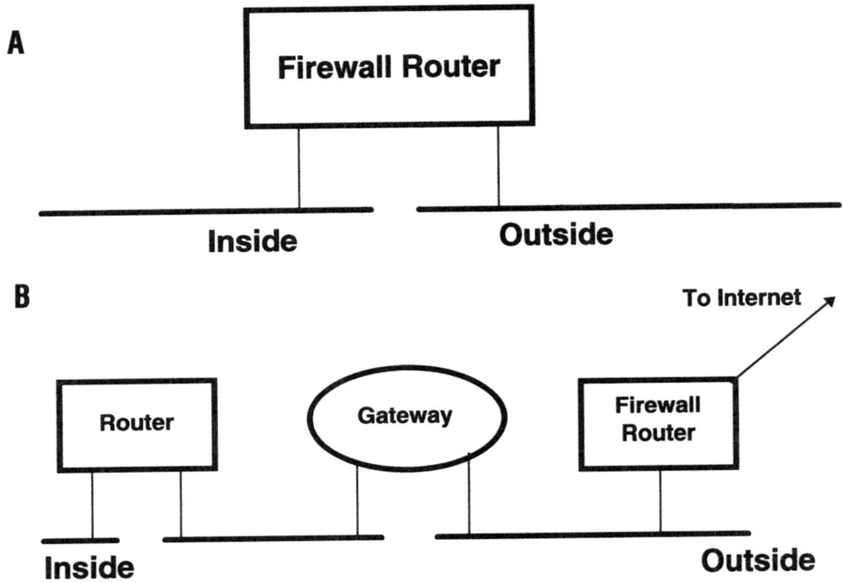

be done so after careful consideration, weighing security risks, needs, and costs.

Forging an Internal Acceptable-Use Policy

Once the security policy at the connection point is met, many issues are raised about an organization's own Internet acceptable-use policy. Will everyone in the organization be allowed a direct Internet connection? Will every Usenet newsgroup be available 24 hours a day? How will training and support regarding secure and prudent use of the Internet connection be handled? These are only a few of the issues that must be dealt with if the Net is to be properly used for the safety of the enterprise. The Intel Corporation, for example, has published an extensive set of guidelines[7] for its internal users, including clear definitions and behavior resulting in disciplinary action, as well as behavior considered to be good etiquette ("netiquette"). Simple but important aspects of netiquette involve issues like the proper use of disclaimers in e-mail and Usenet postings, respecting the posted working hours of FTP sites, and refraining from attempting to establish Telnet sessions on machines with no guest or anonymous account.

Although simple, these clear guidelines set a tone for the corporation's use of the Internet connection, based on reasonable expectations that are the very roots of the Internet Community. As more and more corporations and varied users connect to the Net, this passing on of folklore is more than a good thing. It allows the Net to grow and flourish as a community, instead of becoming a dangerous, urban jungle. Guidelines for each corporation will vary on the basis of corporate goals, sophistication of the user community, and the nature of the corporation's business, but it is imperative that a policy be proposed and available to everyone using the Internet connection.

Costs (Human and Machine)

As was discussed earlier, some out-of-pocket costs are associated with an Internet connection. Large corporations wishing to provide high-speed access to their internal users may need to purchase communications equipment (routers, intelligent hubs, etc.) in addition to their connectivity costs (as high as $50,000 per year). Of course, most companies will wish to maintain and monitor their own connection, which requires trained network personnel. A considerable challenge to companies wishing to provide high-quality Internet support to their customers is the interdisciplinary skill set that is necessary.

A desktop user, for example, may attempt a `Mosaic` WWW connection to an external server halfway around the world. If the connection fails, there could be many sources of error: the TCP/IP stack on the user's desktop computer; a `Mosaic` or operating system parameter; an error on the interface card, local hub, or router; the configuration and status of the local or remote domain name server; a hardware error between the internal and external network; or an interdomain routing problem (among others!). Clearly, personnel in network support, desktop support, network operations, and other functions must work together to diagnose and solve Internet-related problems and be provided the resources to do so. In the new "distributed computing" world of the Internet, the need for communication, collaboration, and constant education also applies, but to the I/T support staff.

User Education, Training, and Survival

Even after a cohesive security policy is established and implemented and support infrastructures are in place, a considerable amount of work remains when considering deployment of Internet-based services across a company. In order to be successful, companies must have a commitment to user training. Internal customers used to Microsoft `Word` or `Lotus 1-2-3` will find a tremendous change with the additions of `Mosaic`, `TurboGopher`, or `Eudora` (PC- and Mac-based e-mail software). The training infrastructure must become both dynamic and flexible. Most likely, toolsets will continue to change at a frantic pace, demanding both effective deployment at the organizational level and much more highly trained users at the local level. At Rohm and Haas Company, for example, our "standard" version of a public-domain VT-100 terminal emulator has changed three times in 1 month. We are investigating ways to more quickly and easily provide updates of the tools while providing a foundation of service.

One simple way of providing access to the latest versions of tools is an internal anonymous FTP server. We provide the newest versions of popular Internet tools from such a server, including on-line documents describing basic installation and configuration options. Our "production" versions of the tools are kept on our local area network (LAN) file servers. This scenario is not perfect, but it allows the more adventurous customers to try new versions of software, while allowing less-experienced customers to use the standard tools.

Rohm and Haas also uses a combination of hardware and software (Digital Equipment Corporation's `PathWorks` and Novell Inc.'s `NetWare`, `Ethernet`, and `Token Ring` and IBM-PC, Apple Macintosh, and UNIX workstations). The heterogeneous nature of our computing environment has forced us to consider carefully when choosing the tools that will be generally available.

The Internet is a great example of Darwin's theory of the "survival of the fittest". Tools that do not work fall by the electronic wayside fairly quickly, and those that do work are constantly being updated and improved. For example, we began our use of the WWW browser `Viola` (written by Pei Wei) in 1992. It is an excellent program, but it was quickly upstaged by `Mosaic` (by Marc Andreesen, Eric Bina, and others from the National Center for Supercomputing Applications or NCSA). The WWW browser functionality soon was available to IBM PCs and Apple Macintoshes, in the form of `WinMosaic/MacMo-`

`saic` (NCSA) and `Cello` (by Thomas Bruce), followed by `WinWeb` and `MacWeb` from Enterprise Integration Technologies (EIT). Most of these tools have gone through extensive revision and improvement cycles over the past 12–18 months, and most are vastly improved from the prerelease code that was available originally.

The evolution of the WWW browser is a good example of the life cycle that Internet software follows. First the tool becomes available in a simple form. If the tool is found to be useful, it is used by hundreds or maybe even thousands of users. Feedback is sent to the developer(s), who frantically try to incorporate all the suggestions (and in some cases, contributed bug fixes or new modules). After a time, "competitors" to the tools spring up and are themselves rigorously tested and either improved or discarded. The "survivors" are those tools that have proven themselves to work reliably under the most demanding circumstances and have shown continuous improvement. Tools that are no longer useful or fall into disrepair are simply discarded. The dynamism of this process improves the tools quickly, but it can become a corporate nightmare for organizations looking to "standardize" on a particular release of a tool. Deploying Internet tools is a balancing act between deploying tools that are robust and stable and using the newest tools that provide the most features.

Learn from the Customer

I/T staffs must also begin to depend on their internal customers to help shape the direction that the local computing environment will take. For example, business people, researchers, and others in the organization must begin to take an active role in specifying both the information resources they need (e.g., Library of Congress resources, National Institute of Health archives, and Federal Register databases) and the tools they prefer (e.g., Windows-based Z39.50/WAIS front-end, UNIX-based MIME mail reader). With the explosion in Internet activity and information that is available, I/T staffs cannot possibly keep up with all the possibilities that global networks offer. Without the help of the customers who need and use the information resources, quite possibly the I/T staff will choose the wrong or inappropriate computing tools. This possibility implies that I/T staffs must work more closely with internal and external customers and actually begin to learn from them.

Historically, I/T staffs have been the providers of hardware and software, drivers of new technology, and custodians of the networking infrastructure. I/T's challenge is to listen to the customers and make the foundational components (security, access, and training) available to them. This new relationship demands flexibility on the part of the support provider and more involvement on the part of the customer.

Paradigm Shifts

The greatest challenge afforded by the Internet is the huge paradigm shift that it presents and represents. Long before the Internet existed, however, another paradigm shift had taken place: that of the personal computer (PC).

More Powerful Than the Telephone—The Real Use of Computers?

Before the introduction of the IBM PC in 1981, most computing was performed on timesharing, mainframe-class computers. Although severely underpowered by today's standards, the IBM PC allowed, for the first time, computing on an individual basis. By 1987, millions of PCs were in business and at home, with an entire industry (software, peripherals, training, and service) to support them. To this add the introduction of Microsoft Windows in 1990, and its ubiquitous presence by 1992. Nearly every successful business around the world uses some form of personal computing for its accounting, word processing, graphics, and e-mail functions. In fact, most businesses would be hard-pressed to operate normally in the absence of the PC. We have truly seen a paradigm shift from centralized timesharing to decentralized personal computing.

The Internet and what it represents is a more dramatic and important paradigm shift than the PC, and indeed the Internet represents the most dramatic paradigm shift since the telephone. This technology allows us to communicate with each other similarly (like using the telephone), but it also allows us to share resources like documents, images, and even computer programs. We can communicate with people anywhere in the world as if they were in the next room. On-line technologies allow us to capture ideas and share them at a later time, as a personal library of sorts. Documents that are important to us today may become important to others tomorrow. The Internet allows us to communicate and share information through time and space.

The industrial use of this technology has also grown. The ARPANET became a production network in 1975. By the early 1990s millions of computers belonged to the global Internet. But only in the past few years has the ubiquity of Internet access made global e-mail useful. FTP, Gopher, and WWW services allow suppliers and individuals to share multimedia information in ways never before considered. The rise of commercial Internet information services (e.g., Dialog, CommerceNet, NetMarket, and MecklerWeb) has redefined the meaning of marketing and customer service. Customers can search suppliers' servers for answers to technical questions, product and contact information, and latest versions of software and journals. Communications are more customer-oriented than ever before, as the customers access the supplier at their convenience, for the information they are interested in.

The Net has also caused corporations to look externally for expertise and diverse opinions, something heretofore unheard of. For example, I/T support people can now scan Usenet newsgroups to gather market research about existing or emerging products and thereby save much time, effort, and money. Often, actual users of software, hardware, or services are willing to share their experiences with prospective customers. This sharing gives the potential customer better data on which to base decisions and encourages vendors to read the newsgroups, remain in electronic contact with their customers, and improve their products and services.

The PC was the beginning of user-centered computing, and the Internet extends that immediacy into a global community. Desktop tools, once limited to serving the needs of isolated PC users, now can interact with other information resources anywhere in the world. Truly, this is personal computing with global reach.

Everyone Is an Information Provider

The other major paradigm shift is a combination of the growth and maturation of Internet technology and the rise of decentralized computing that began with the advent of LAN-based computing. In the 1970s and 1980s, information users were forced to deal with centralized MIS groups to negotiate for computer resources, access to data, and the creation of complex information systems. This situation led to the proverbial "glass house" MIS department, hopelessly trying to meet the dynamic needs of its customers but faced with the long life cycles of mainframe-based software development.

By the mid-1980s, it was clear that moving to commodity hardware (PCs), PC-based file servers, and client–server development tools was the most cost-effective way to deliver computing resources to local customers. However, most of these strategies add wide-area network capability as an afterthought, and many are built on proprietary software standards. Most client–server development methodologies are not designed for the casual user to develop code with. For all the progress, customers' information may still be held hostage by the (now decentralized) MIS department.

Enter the Internet, and tools such as the World Wide Web. The Web changes the rules, in that everyone can easily and cheaply become an information provider, including the researchers in the lab, the lawyers in the legal department, and the office user wishing to share information with the rest of the organization (or the rest of the world, for that matter). The paradigm shift is simply this: No longer are information users simply consumers; they are producers as well.

The Web, built from the ground up using open systems (HyperText markup language, or HTML, and HyperText transfer protocol, or HTTP) and an open standards process, was partially developed by its user community. It is a simple matter to create one's own WWW server and begin publishing information to an entire company (or the entire WWW Internet community) without much assistance from the I/T support group. The use of Internet-based information-sharing tools (including WWW, Gopher, and WAIS) has considerable potential to finally begin to "pay off" on the promise of increased productivity and improved communication promised to computer users so many years ago. We are beginning to see a true web of information being owned, maintained, and shared by those who produce and need the information, and the I/T staff becoming true facilitators of information sharing, instead of the "gatekeepers" of information they do not understand, need, or maintain. The Internet then becomes an agent of change, not because of the information contained within it, but because of the mechanisms it uses and the lessons it teaches. The Internet is closer to the new frontier than the mine field, even given the considerable challenges it presents.

Tea Leaves and Entrails: Predictions

Any prediction about the near-term destination of the Internet will be, at best, a wild guess. A number of imminent developments give us a few clues, however:

- The TCP/IP communications stack is fast becoming a commodity item (e.g., Microsoft will bundle it in their upcoming Windows 95 product, and Novell's TCP/IP is virtually free).
- Internet tools are becoming commercialized (`SpyGlass`, O'Reilly's `Internet-in-a-Box`, Netscape Communications Corporation's `Netscape`, IBM's `WebExplorer`, etc.).
- Point-to-point protocol (PPP)[8] is gaining wide acceptance, allowing mobile users to access remote resources as though they resided locally.
- The number and quality of chemistry-, business-, and government-related Internet resources are continuing to grow at a breakneck pace.

In the near-term, the expansion of Internet access from office and home desktops will continue to grow as interest increases. Wide-area communications must also become easier to set up for home users. ISDN may become important to the home user as a high-speed alternative to dial-up modems. As in the past, faster bandwidth to the home or office leads to more use and more demanding network applications. This cycle of increased available bandwidth and increased need for bandwidth will continue and will spur the demand for improved telecommunication infrastructures and simpler setup and operation.

This point is important: For Internet and related technologies to continue to become important, the underlying connectivity issues must become more transparent to the user. With initiatives such as PPP and the dynamic host configuration protocol (DHCP)[9], users soon will be able to request an Internet connection (at home, office, or on the road) in much the same way they request a new telephone installation. This ability, coupled with improved (and commercialized) versions of Internet tools should contribute to strong growth and sustained interest in the Internet. For the chemistry audience, this interest will continue to build the virtual worldwide chemical community and foster closer work between scientists who are geographically dispersed.

Longer-term, there are even more exciting possibilities. Present-day developments with long-term impact include

- Companies like NetMarket,[10] CommerceNet,[11] and Netscape Communications[12] are offering encrypted transaction services using the WWW.
- Small and large companies are building an Internet "presence" offering commercial products, direct ordering, and product information for individuals and corporations.
- Intelligent software agents, used to automatically search for resources that a user is interested in, are being discussed.
- The World Wide Web Organization (W3O) is being formed, a joint venture between CERN, Switzerland (the inventors of the WWW) and MIT. Tim Berners-Lee (the inventor of the WWW) will be the director.

The WWW effort is based on very simple principles: simple, standards-based protocols working to complete small, fast transactions. The client takes raw information and processes it, perhaps invoking a graphics viewer and sound player, or opening another remote connection. This is the foundation upon which many sophisticated services are built. A good example is NetMarket's acceptance of secure transactions using the PGP (pretty good privacy) encryption system.[13] This was the first popular implementation of secured transactions using the Internet. Secured transactions between trading partners using the Net will become routine, as this must be the case before commercialization can effectively happen.

Most companies (large and small) will create an Internet "presence". This presence will afford companies trust, repeat business, and a credibility that nonconnected companies will lack. The depth and quality of the on-line information available to customers and the speed of service (e-mail turnaround time, reliability of information services, etc.) will differentiate good companies from poor ones.

To quote Michael L. Dertouzos,[14] the director of the MIT Laboratory for Computer Science:

> [The WWW] has a great deal of this information market idea in it already. It's got two beautiful things. The first is an obvious one, that anyone can produce something. You can create a home page that's as elaborate or as silly as you want. The second part of this is that there are 30 million of you each doing this, and it is growing into a beautiful edifice. That's how it will continue to grow: If you do a great Web demo for a health

center, then every health center and clinic is going to want to be "Webbed". That's how it will work.

CommerceNet, W3O, and other consortia that merge the strengths of the commercial and academic communities will further the capabilities of world-wide information exchange and electronic commerce.

In the more distant future, the creativity that has marked the advances in the Internet and information technology will continue and even escalate. New information-based products will present themselves, from personalized "multimedia-papers" (newspaper-like entities with text and graphics, live video, on-demand advertising, multiway voice channels, etc.) to true advances in the way information is sorted, stored, and presented. Currently, the metaphors we use (book pages, file cabinets, mailboxes, etc.) are adequate for the limited types of information we deal with every day, but they are clumsy in their integration and organization. The larger issue to deal with is the ways people actually use information, once they find it. Classical I/T is preoccupied with the creation, movement, and management of data (not information). Future challenges will center around human–machine processing of information and its application to improve our lives, both personal and corporate.

As Yogi Berra observed, in the midst of change, the future isn't what it used to be.

References

1. Krol, E. *The Whole Internet User's Guide and Catalog;* O'Reilly & Associates: Sebastopol, CA, 1992, p 11.
2. "About the Commercial Internet Exchange", accessible via WWW at URL **http://www.cix.org/about-cix.html**
3. *The National Information Infrastructure: Agenda for Action;* National Telecommunications and Information Administration (NTIA), Department of Commerce: Washington, DC, 1993.
4. Bright, M.; Dyott, T. M.; Fujimoto, T. T.; Pierce, T. H.; Reynolds, C. H.; Stuper, A. J.; Zander, G. S. "MOLY-86: An Interactive Molecular Modeling System", *QCPE Bulletin* **1987,** *7,* 2.
5. Cheswick, William R.; Bellovin, Steven M. *Firewalls and Internet Security;* Addison–Wesley: Reading, MA, 1994.

[6]Hambridge, Sally; Sedayao, Jeffrey C. "Horses and Barn Doors: Evolution of Corporate Guidelines for Internet Usage", Intel Corporation, originally published in the *USENIX Proceedings of the Seventh System Administration Conference;* November 1993. Also accessible via WWW at URL **http://www.jf.intel.com/about-intel/papers/horses.html**
[7]"Intel's Internet Guidelines", accessible via WWW at URL **http://www.intel.com/about-intel/papers/intel_guidelines.html** and **http://www.intel.com/intel/papers/intel_guidelines.html**
[8]Simpson, W. "The Point-to-Point Protocol (PPP)".
[9]Alexander, S.; Droms, R. "DHCP Options and BOOTP Vendor Extensions".
[10]NetMarket.com, accessible via WWW at URL **http://www.netmarket.com**
[11]CommerceNet, accessible via WWW at URL **http://www.commerce.net**
[12]Netscape Communications Corporation, accessible via WWW at URL **http://home.mcom.com**
[13]Zimmerman, Philip "Pretty Good Privacy (PGP)", information available via FTP at **net-dist.mit.edu/pub/PGP**
[14]Denison, D. C. "Spiders and Shovels: Michael L. Dertouzos on W3O", O'Reilly & Associates: Sebastopol, CA. Published as part of O'Reilly's Global Network Navigator (GNN), and accessible via WWW at URL **http://nearnet.nnc.om/news/feature/mit.cern/dertouzos.html**

Appendix: Data from the External Survey

This is a brief statistical breakdown of the data from the external survey.

- Number of responses: 65
- Percentage of industry responses: 58% (38)

Current Uses of the Internet by Respondents

Percentage (Frequency)	*Use*
54%* (35)	Communications using e-mail to and from industry and academia
37%* (24)	Accessing bulletin boards and mailing lists
34% (22)	Performing collaborative research
29% (19)	Obtaining files and software via FTP
22% (14)	Performing on-line library, catalog, and database searches (Dialog, Chemical Abstracts, etc.)

15% (10)	Using Internet discovery tools to obtain software, documentation, etc.
11% (7)	Obtaining support, software fixes, and license keys from suppliers and vendors
11% (7)	Using remote computer power
8% (5)	Human resource issues (choosing a company on the basis of Internet access, obtaining alumni information, etc.)

*This data reflects the number of respondents that specifically mentioned the particular item. In reality, all the respondents used e-mail, bulletin boards, or mailing lists to read and reply to the posting.

Major Concerns of Respondents Regarding Internet Usage

Percentage (Frequency)	*Concerns*
31% (20)	Security (political issues, secrecy, break-ins, etc.)
8% (5)	Difficulty for novices to get started and find resources
6% (4)	Time "sink"
3% (2)	Reliability
2% (1)	Uncompressing and manipulating files received via FTP
2% (1)	Difficult to manage LAN–WAN software on PCs

Respondents

Industrial Organizations: 3M Graphic Research Lab; Akzo, The Netherlands; BP America; Burroughs Wellcome Co.; Cambridge Entrepreneurial Network; Chemical Concepts Corporation; Chemistry Editorial—Heidelberg; Chevron Research and Technology Co.; Cray Computer; DuPont; Eastman Kodak R&D; Exxon Research and Engineering; Ford Electronics; Genentech, Inc.; General Electric (Aircraft Engine Division); Genetics Institute; Glaxo; IBM; ICI Pharmaceuticals; MMC; MSEN; Monsanto; PPG; Pfizer Central Research, U.K.; Sandoz Research Institute; SmithKline Beecham; Stam Tijdschriften, The Netherlands; Sterling-Winthrop; Thinking Machines Corporation; Union Carbide; Wavefunction; Westinghouse Electric Corporation.

Academic and Governmental Institutions: Bijvoet Center for Biomolecular Research, University of Utrecht; Bradley University; City

University of New York; Columbia University; Connecticut State University; Drexel University; Duke University; Kent State University Libraries and Media Services; Notre Dame University; Ohio State University; Massachusetts Institute of Technology; Mississippi State University; National Science Foundation; University of British Columbia; University of California, San Diego, University of Texas at Austin; University of Texas at Dallas; Washington University Center for Molecular Design; Wayne University; The Wistar Institute of Anatomy and Biology, University of Pennsylvania.

appendix

Anonymous FTP

Steven M. Bachrach

FOR THOSE WHO HAVE NEVER USED ANONYMOUS FTP, we present a quick primer here. FTP stands for file-transfer protocol and is a set of instructions on how files are sent between computers on the Internet. FTP is also the name of the program that sends files across the Internet and is what we will discuss here.

To obtain or send a file to another computer you will need an FTP program and the Internet address of the target computer. There are a number of different FTP packages; one should be included within any Internet start-up package, and FTP is included within any UNIX operating system. Most FTP packages operate in a very similar fashion, so the instructions here should be applicable to your configuration.

To start the FTP session, at the system prompt, type the command

```
ftp target.computer.address
```

where `target.computer.address` is the address of the computer that you wish to communicate with. The target computer will then respond with a request for an account name. Most often you will not have an account on the target computer, and therefore you will not be able to communicate

3223–7/96/0309$06.00/1

with that machine. However, many software and information providers have established an anonymous FTP account enabling anyone to obtain files from that computer. This is what is commonly referred to as *anonymous FTP*. To log in, you would then type

```
anonymous
```

and the target computer then will prompt you for a password, usually requesting that you enter your electronic-mail address as the password. This request is made so that the administrator can monitor the usage of the anonymous FTP site.

Once the password (your electronic-mail address) has been entered, the target computer will respond that the FTP session has begun and will present the standard FTP prompt:

```
ftp>
```

You are now free to access the files and directories set up for that anonymous FTP account One traverses the directories using the standard UNIX commands. The most important are **pwd** (present working directory), which presents the name of the current directory; **ls** (list), which provides a list of all files and subdirectories contained within the current directory; and **cd** (change directory) followed by a subdirectory name, which moves you into that subdirectory. Once you have navigated to the correct directory and located the files of interest, you can proceed to transfer the files to your computer, which is sometimes called **downloading** a file.

- The command to transfer a single file is

  ```
  get filename
  ```

  where `filename` is the name of the file you wish to download. The file will be placed in your current local directory with the same file name as it had on the target computer.

- To obtain many files at once, use the command

  ```
  mget filenames
  ```

  where usually `filenames` is a truncated name with a wild card (a character such as an asterisk) appended. For example, to obtain the files Test10, Test20, and Test30, you could use the command `mget Test*`. The target computer will then prompt you for transfer of each of the files available.

- To send a file to another machine, the command is

  ```
  send filename
  ```

- To send more than one file at a time, use the command

  ```
  mput filenames
  ```

  where again you can use a wild card name to group all the files.

- To terminate the FTP session, type the command `bye`.

Two aspects of the FTP procedure are slightly tricky. The first is that you must determine the location that the files will be placed on your computer. Many programs have a menu selection to designate the directory in which to place the transferred files. The other option is to use the command

```
lcd directoryname
```

at the FTP prompt, which changes the local directory name and will place the downloaded files into that directory.

The second tricky issue is the format of the data that will be transferred. Files can be stored in ASCII or binary format. Plain text files should be transferred in ASCII format. Programs and compressed data should be transferred in binary format. At the prompt, set the file type to ASCII with the command `ascii` or to binary with the command `binary`.

These instructions are really just an introduction to the power of this utility, but is sufficient to get you started. More information is available by entering the command `help` at the ftp prompt when logged into an FTP session.

Finally, the World Wide Web browsers have a built-in ability to access FTP sites. You simply enter the location of the FTP site as a URL (uniform resource locator). For FTP sites, the URL begins **ftp://**, as in the example for the NCSA (National Center for Supercomputing Applications) anonymous FTP site: **ftp://ftp.ncsa.uiuc.edu**. Navigation through the site and selection of files is accomplished by clicking the mouse on the desired locations or files. The client handles the file type selection, and you will be prompted for the location to save the files.

Glossary

Steven M. Bachrach

anonymous FTP One of the major drawbacks to the standard FTP session is the need for an account with a password on the target computer. The alternative is to establish an anonymous account on the target computer to which anyone will have access. The usual procedure is to set up the anonymous account with the password as the account name of the incoming user. For example, if Bill Clinton wishes to obtain *Mosaic* from the National Center for Supercomputing Applications (NCSA) using FTP, he would type the following commands (his are shown in italics, the computer's responses are shown in plain text):

```
ftp ftp.ncsa.uiuc.edu
name: anonymous
Guest login ok, send ident as password
password: president@whitehouse.gov
```

Establishing the anonymous FTP account is relatively straightforward and is usually explained in the manual (man) pages for FTP. The only concern is that one needs to be careful so as to restrict access to the computer. The anonymous FTP account can be a security leak. Further details are given in the appendix on page 309.

3223–7/96/0313$12.00/1

Archie A software tool for searching FTP sites, developed at McGill University. Archie gathers and indexes lists of files available on anonymous FTP sites. Archie periodically polls all the FTP sites it knows of and determines all the files available. A user then connects to one of the sites running Archie and makes a search request. The simplest way to perform an Archie search is to telnet to one of these sites. Each site will have help available. A search is initiated by the command

```
prog filename
```

Archie will then return a list of all FTP sites that match that filename. Other ways of performing Archie searches are given in more general Internet how-to books. Some Archie sites are

archie.rutgers.edu (northeast United States)
archie.sura.net (southeast United States)
archie.unl.edu (west United States)
archie.mcgill.ca (Canada)
archie.au (Australia)
archie.funet.fi (Europe)
archie.doc.ic.ac.uk (United Kingdom)

ATM or **asynchronous transfer mode** A protocol for transmitting data across a network. ATM allows for very rapid transfer speeds and probably will play a major role in future networks.

Bitnet One of the networks that is part of the Internet.

bookmark A place marker for the address of a Gopher site. Instead of having to remember the address or the path one took to find a site, the Gopher client can place the address in a bookmark file. The user can then simply select the entry from the bookmark file to return to the site the next time. The bookmark file is also called a hotlist.

browser Software that reads text marked with HyperText markup language (HTML) so that you can access World Wide Web sites. Examples are `Mosaic`, `Netscape`, and `Lynx`. Also called client.

CERN or **Centre Européen pour la Récherche Nucleaire** A physics particle accelerator laboratory in Geneva, Switzerland, where the World Wide Web originated.

client Software used to obtain data from a server. Browsers are an example of clients that access data from WWW servers.

discussion group A collection of individuals who are interested in exchanging ideas on a specific topic via electronic mail. Individuals subscribe (or register) with the group to participate.

download To transfer a file from a remote computer to your own computer.

electronic mail or **e-mail** Messages sent from a user on a computer to another user who has an account on any computer on the network. Electronic-mail messages contain a number of fields, typically a `From:` line, indicating the address of the author, and a `Subject:` line, which gives a brief indication of the enclosed message. Messages usually contain text, but files can be attached. Many users place a signature at the bottom of their messages that contains their name, address, phone number, etc. Many mail programs simplify reading and writing e-mail messages. More detail is given in Chapter 2.

e-mail Electronic mail.

e-mail shorthand Abbreviations and other conventions often used in e-mail. Examples of abbreviations are BTW, by the way; WRT, with respect to; IMHO, in my humble opinion; FWIW, for what it's worth; OTOH, on the other hand; and RTFM, read the freaking manual. Some conventions are the use of asterisks or underscores before and after a word for emphasis (because bold or italic is not available) and the use of sideways smiley faces :-).

Ethernet A protocol for a local network using, at first, coaxial cable. Over time the wiring guidelines for Ethernet have changed to include thin coaxial cable and 10-baseT, which is similar to telephone cable. What has not changed significantly is the way messages are sent over the wiring according to the Ethernet protocol. This protocol specifies how to create the message (using packets), how to define addresses, and how to propagate the packets. Typical data-transfer speed in Ethernet is about 10 Megabits per second.

FAQ or **frequently asked question** Discussion groups are often plagued with repeated posting of the same questions, particularly as new members join the group. To minimize these repeated questions, many discussion lists produce a file containing the most frequently asked questions along with their answers. These FAQs are then made available to all, usually through an anonymous FTP site associated with the discussion group.

firewall A computer system set up on a local network to preclude access to other systems; that is, a firewall keeps a local network separate. For example, a company may wish to allow its employees to have access to the Internet, but not allow outsiders to have access to the company's computers. A firewall is used to prepare just such an environment.

freeware Software made available at no cost by the author. Even though the program is available for free, many authors have copyrighted their work; the software may therefore be protected and its distribution limited.

FTP or **file-transfer protocol** The basic protocol that defines how computers transfer files between themselves. FTP can also refer to a program that facilitates the transfer of files using the file-transfer protocol. Normal FTP operation requires that the user know the address of the target computer, have an account on the target, and know the locations of the files of interest. Specific commands allow the user to connect to the target computer, navigate through the directory structure, and then send or receive files. The FTP program is standard on all UNIX systems and can be obtained in a variety of packages for PCs and Macintosh computers. The appendix Anonymous FTP gives specific instructions.

FTP site A computer that stores documents that may be accessed through FTP.

gateway A computer that connects a local network to the Internet.

GIF or **graphics interchange format** A standard format for storing image files by digitizing and compressing graphic images.

Gopher A program developed in an effort to streamline and simplify use of the enormous resources available on the Internet. Gopher allows users to obtain data without knowledge of FTP commands or addresses. Gopher is actually two pieces of software, a server and a client. The Gopher server provides access to files on the server computer and transparent links to other Gopher servers. The Gopher client connects the user to the server and allows the user to display data and navigate through GopherSpace. Gopher is limited to supplying normal text files, links to other Gopher servers, WAIS searches, Telnet sessions, and audio files. Gopher is a menu-driven navigator of the Internet. Gopher is available via anonymous FTP to **boombox.micro.umn.edu**. Chapter 5 gives further details on Gopher.

GopherSpace The network of all Gopher servers. It encompasses all data and information served on all Gopher servers. The University of Minnesota maintains a fairly comprehensive list of all Gopher servers and links to these servers; therefore, in principle, a user can traverse all Gopher servers and their files.

helper applications Software that plays video or sound or that processes some image formats. This software is used by browsers but is not included in browser software in order to keep the developers' costs low and to keep the browser small.

home page The opening page of a World Wide Web site.

hotlist *See* bookmark.

HTML or **HyperText markup language** A protocol for coding text with tags so that files can be integrated into the World Wide Web. HTML describes how to create text in different fonts, sizes, and styles; how to create a link to another file or server; how to imbed graphics; how to create forms; etc. HTML-coded text is called HyperText. HTML is a type of SGML (standard generalized markup language).

HTTP or **HyperText transport protocol** Defines how World Wide Web servers provide their files and how one connects to these servers.

HyperText A type of file that allows for direct links to other files when the user selects keywords in the document. The best way to think of this is by analogy. Suppose you have a large encyclopedia at hand and you are looking up the entry on Great Britain. You come across the mention of Winston Churchill and decide you would like to read more about him. In traditional forms, you would have to get the "C" volume of the encyclopedia and find the Churchill entry. In HyperText, Churchill's name in the Great Britain entry would be highlighted as a keyword. Selecting this keyword would automatically bring up the Churchill entry. The Churchill entry may point out that Churchill's hobby was painting and highlight the word painting. Selecting this keyword might bring an image of one of his more brilliant works. Another link in the Churchill entry might be to an audio recording of one of his speeches. This keyword selecting can go on indefinitely.

IMHO or **in my humble opinion** e-mail shorthand. When answering an e-mail or a post on a discussion group, the author may expound on his or her opinion at great length, and so to somewhat diminish the pompous effect, the abbreviation IMHO is appended.

in-line image An image that is merged with text.

Internet A matrix of networks and computer systems linked together around the world. Also called the Net.

Internet protocol (IP) address A unique address assigned to every host computer on the Internet. The address is composed of four numbers separated by periods, such as **aaa.bbb.ccc.ddd**. People handle words easier than they do numbers, particularly in terms of remembering locations, and so the address can also be expressed as (up to) four words separated by periods. For example, the IP address for the Northern Illinois University Chemistry Gopher Site is **131.156.9.16**, and its name is **hackberry.chem.niu.edu**.

Internet provider A company through which you get connected to the Internet. You also need software such as PPP or SLIP.

ISDN or **Integrated Services Digital Network** Digital telephone systems. For computer networks, the advantages of ISDN are primarily twofold. First, data can be transferred at much higher speeds than through regular phone lines, from 56 to 128 kilobytes per second (kbps). Second, ISDN lines can handle simultaneous voice and data transmissions, and they thereby remove the need for a dedicated voice line and a dedicated data line.

JPEG or **Joint Photographic Experts Group** A format for storing images in files; a standard for graphics compression that handles both full-color and gray-scale, but not black-and-white, figures or motion pictures. This format is ideal for compression of large color images, particularly scanned images. JPEG allows for higher quality images than GIF.

LAN or **local area network** A connection of computers, terminals, printers, etc., that is geographically limited, for example, to one organization or even to one department within an organization.

link A connection between one HyperText document and another.

list server Software used to run and maintain a discussion group. The list server acts as a central receiving point of mail, which it then forwards to all members of the group. List servers are sometimes called mail exploders.

`Listserv` A specific list server, the most common one on the Internet.

`Lynx` A popular World Wide Web browser that reads and displays only ASCII text.

mail exploder *See* list server.

mailing list A list of electronic-mail addresses to which messages are sent.

man pages or **manual pages** A full set of on-line manuals included with the UNIX operating system. To access these pages, at the system prompt type `man term`, where `term` is the instruction for which you wish to find further details. For example, to find information about copying files, you would type `man cp`.

MIME or **multipurpose Internet mail extension** A method for classifying the type of file that is transmitted. A graphics file can be sent in a variety of formats, such as GIF or JPEG, each of which has a unique MIME, such as **image/gif** or **image/jpeg**. Through the use of appropriate programming structure, files can be directed to the appropriate reader. For example, on a UNIX workstation, *`Mosaic`* uses a mailcap file to indicate what readers should be invoked for the various types of files it comes across. The standard setting for the mailcap entry for GIF files is to invoke the reader program called *`xv`*. When a user of *`Mosaic`* elects to view an image in a file in the GIF format (the users would most likely not even know what the format of the image is), the file is transferred to the user's computer, the MIME is read (**image/gif**), *`Mosaic`* then uses the mailcap file to determine what reader to invoke (*`xv`* in this case), that application is automatically launched, and the file is presented. As stated before, all of this magic is invoked transparently to the user.

mirror site A site that contains the same files as another, very popular site, set up to relieve heavy traffic at the popular site.

Mosaic A World Wide Web client. Developed at the National Center for Supercomputing Applications (NCSA), *Mosaic* provides a seamless graphical user interface to navigate the Web. When used in conjunction with other programs, *Mosaic* can present HyperText files, graphic items, audio clips, and movies. The program is available for a variety of computers, including many UNIX workstations and Macintosh and Windows PCs. *Mosaic* is available via anonymous FTP from **ftp.ncsa. uiuc.edu** in the **/pub/web directory**.

MPEG or **Moving Picture Experts Group** A protocol for standardizing the digitization of motion pictures.

NCSA or **National Center for Supercomputing Applications** An outstanding resource for a variety of network and software solutions, the NCSA is a federally funded institution dedicated to serving the computing community. NCSA provides, for free, a variety of exceptionally useful tools including *Mosaic*, Telnet for the Macintosh, and graphics programs. NCSA can be reached via anonymous FTP to **ftp.ncsa.uiuc.edu**, via Gopher at **gopher. ncsa.ncsa.uiuc. edu**, and via the World Wide Web at **www.ncsa. uiuc.edu**.

netiquette Network etiquette; guidelines for good manners on the Internet.

Netscape A popular new browser for the World Wide Web, *Netscape* incorporates many of the features of HyperText markup language (HTML) version 3.0, which includes tables. *Netscape* is the product of Netscape Communications, which also markets a World Wide Web server.

newsgroup A discussion group on a specific topic, usually part of the Usenet News bulletin board system.

post An e-mail message.

PostScript A language for describing the appearance of images on a page. PostScript was developed by Adobe Systems as a means for uniformly describing a page for a variety of sources, such as a screen and hard-copy devices.

PPP or **point-to-point protocol** The fastest growing protocol for connecting individual users to the Internet over telephone lines.

RFC or **request for comments** A truly remarkable mechanism for proposing and establishing standards and protocols on the Internet, the RFC is a note sent out to the Internet community for general discussion. The note contains an individual's (or a small group's) suggestions and recommendations on setting a standard for a particular computer or network operation. These notes are debated until a consensus is reached. What is most amazing is that these RFCs are then generally accepted by vendors, software suppliers, and freeware authors and serve to guide the generation of useful computing environments. The RFCs are numbered and continue to be referred to as RFCs even after they have been adopted as the standards. RFCs have set the standard for the handling of e-mail, MIME, HTML, and other aspects of the Internet discussed in this book. (For amusement, imagine a corporation or academic unit coming to an agreement in the same manner as RFCs are produced.)

router A device for transmitting messages between networks.

RTFM or **read the freaking manual** A curt response to particularly useless (to a newsgroup) questions that are easily answered by simply reading the appropriate manuals. A useful warning is thus offered: post a question only *after* you have tried to obtain an answer on you own and failed. Discussion groups are not replacements for libraries, manuals, or some effort on your own part.

server A program running on a computer that is linked to a network. It responds to requests from client programs.

SGML or **standard generalized markup language** A system for defining markup languages and documents. HTML is one kind of SGML.

shareware Software distributed for demonstration purposes, but if you decide it is useful, you are expected to send the author a small fee, typically less than $25. A great deal of extremely useful software is shareware, and shareware authors should be encouraged to continue to provide their products in this manner.

SLIP or **serial-line Internet protocol** An older protocol for connecting computers to the Internet over telephone lines, often using a modem. SLIP is slowly being superseded by PPP.

TCP/IP or **transmission control protocol/Internet protocol** The standard protocol for Ethernet transmission, adopted by all UNIX suppliers. This is actually two sets of protocols (TCP and IP), but they are usually lumped together to address all the major communications needs, such as FTP and Telnet.

Telnet The Internet protocol for remote terminal connection to another computer, using TCP/IP. Telnet allows the user to remotely log on to another computer. A Telnet session is activated by entering the Telnet command followed by the address of the remote computer.

thread A series of posts (or e-mail messages) on the same subject.

URL or **uniform resource locator** A brief description of the location of any file, server, or other resource on the Internet. The format of a URL is

scheme://host.domain[:port]/path/filename

The scheme tells your browser how you are accessing the information, such as gopher, ftp, http, wais, news, or telnet. The host.domain is the Internet address of the server to be accessed, that is, the site where the information is stored. The path and filename together are the location of the information on the host computer.

Usenet A large group of networks and computers organized by newsgroup.

Veronica (Very Easy Rodent-Oriented Net-wide Index to Computerized Archives). A simple-to-use database of GopherSpace that is accessed through a Gopher client. A relative of Archie, Veronica supplies a similar service for Gopher servers as Archie does for FTP sites. Veronica creates a database of directories and files available on Gopher servers that are registered with it. Veronica can be accessed through the Gopher link:

```
Name=Veronica
Type=1
Port=70
Path=1/veronica
Host=futique.scs.unr.edu
```

W^3 *See* World Wide Web.

WAIS or **Wide Area Information Service** A text-based search engine developed at Thinking Machines, Inc. The WAIS program creates an index of every word in a collection of files. Users can then search for any word within this collection of files. WAIS locates all the files that contain the search term and presents a list of the file names, ranked in order of the frequency of appearance and location of the term within the file. WAIS can be integrated within Gopher, providing a simple front-end interface to the WAIS database. Otherwise, one must use the WAIS client. WAIS is currently available as freeware.

WAN or **wide-area network** A network of connected computers, terminals, printers, etc., that extends over a large geographical region, typically more than a 1-km radius.

Web *See* World Wide Web.

Web site A World Wide Web site. A computer that contains files that may be accessed through the World Wide Web.

World Wide Web The World Wide Web (WWW or W^3 or Web for short) is a HyperText Internet navigation system. It is really a HyperText analogue of Gopher. Instead of needing to know Internet addresses, FTP commands, etc. (which are needed for anonymous FTP) or navigating by menus (as is done in Gopher), WWW leads you through the Internet using HyperText links. WWW supports links to text, graphics, audio, and movie files; HyperText files; anonymous FTP sites; WAIS servers; Telnet sessions; and Gopher servers. Just like Gopher, the user needs a Web client (of which there are many, *Mosaic* being the most popular) and connects to a Web server. The Web server provides the files and links to other sites. Further details are given in Chapter 6.

WWW *See* World Wide Web.

Bibliography

WE HOPE THAT THIS BOOK will provide answers to most of your questions, but we realize that no book can be all-encompassing. Furthermore, the Internet is constantly changing, and new resources will certainly become available after the printing of this book. As an aid to find further information, we list some popular books on the Internet and some on-line resources for assistance.

Books

December, J.; Randall, N. *The World-Wide Web Unleashed;* Sarrs Publishing, 1994.

Estrada, Susan *Connecting to the Internet, An O'Reilly Buyer's Guide;* O'Reilly & Associates: Sebastopol, CA, 1993.

Ford, A. *Spinning the World-Wide Web;* Van Nostrand: New York, 1995.

Glister, P. *The Internet Navigator;* Wiley: New York, 1993.

Hahn, H.; Stout, R. *The Internet Yellow Pages;* Osburne/McGraw-Hill: New York, 1994.

Keboe, B. P. *Zen and the Art of the Internet;* Prentice-Hall: New York, 1994.

Kent, P. *Complete Idiot's Guide to the Internet,* 2nd ed.; Alpha Books, 1994.
Krol, E.; Loukide, M. *The Whole Internet User's Guide and Catalog*, 2nd ed.; O'Reilly & Associates: Sebastopol, CA, 1994.
Levine, J. *MORE Internet for Dummies;* IDG Books, San Mateo, CA, 1994.
Marine, April; Kirkpatrick, Susan; Neou, Vivian; Ward, Carol *Internet: Getting Started;* PTR Prentice Hall: New York, 1993.
Wiggins, R. *The Internet for Everyone: A Guide for Users and Providers;* McGraw-Hill: New York, 1994.

On-line Resources (Listed as URLs)

Directories of Chemistry Resources

Bachrach, S. "Chemistry Information on the Internet", **http://hackberry.chem.niu.edu/cheminf.html**
Kopeltvich, M. "The World Wide Web Virtual Library: Chemistry", **http://www.chem.ucla.edu/chempointer.hmtl**
Sanios, J. "WebChemistry", **http://www.latrobe.edu.au/chejs/chem.html**
Wiggins, G. "Some Chemistry Resources on the Internet", **http://www.rpi.edu/dept/chem/cheminfo/chemres.hmtl**

Directories of General Internet Resources

"CUI WWW FAQ and Guide", **http://cuiwww.unige.ch/OSG/FAQ/www.html**
December, J. "Internet Tools Summary", **http://www.pri.edu/Internet/Guides/decrmj/internet-tools.html**
"EINet Galaxy", **http://galaxy.einet.net/galaxy.html**
Grange, N. "The Guide to Network Resource Tools", **http://www.earn.net/grnt/notice.html**
"Internet Resource Metamap", **http://www.ncsa.uiuc.edu/SDG/Software/Mosaic/Demo/metamap.html**
Usenet FAQs, **http://www.cis.ohio-state.edu/hypertext/faq/usent/FAQ-List.html**
"What's New with NCSA Mosaic", **http://www.ncsa.uiuc.edu/SDG/Software/Mosaic/Docs/ whats-new.html**

Index

F

V

W

Copy editing and indexing by Janet S. Dodd
Page design by Melinda Grosser
Page layout and additional design by Betsy Kulamer
Cover design by Amy Hayes

Printed and bound by Maple Press, York, PA